The Story of
INTERFERON
The Ups and Downs in the Life of a Scientist

The Story of
INTERFERON
The Ups and Downs in the Life of a Scientist

K Cantell
Finland

World Scientific
Singapore • New Jersey • London • Hong Kong

Published by

World Scientific Publishing Co. Pte. Ltd.

P O Box 128, Farrer Road, Singapore 912805

USA office: Suite 1B, 1060 Main Street, River Edge, NJ 07661

UK office: 57 Shelton Street, Covent Garden, London WC2H 9HE

British Library Cataloguing-in-Publication Data
A catalogue record for this book is available from the British Library.

First published in Finnish by Werner Söderström Osakeyhtiö in 1993

THE STORY OF INTERFERON:
THE UPS AND DOWNS IN THE LIFE OF A SCIENTIST

ISBN 981-02-3148-2

Printed in Singapore by Eurasia Press Pte Ltd

TO AILA

Acknowledgements

I received much wise advice and encouragement from Anto Leikola and Hannu Tarmio and I am grateful to them. These have improved the book and accelerated its publication.

Many of the photographs are from the meetings of the Wetterkulla Medical Center. I remember my friends of Wetterkulla, also the deceased ones, with gratitude for many fruitful discussions and happy times during the past 30 years. Interferon has brought some good friends into my life. One of them is Norman Finter, a pioneer of interferon research, with whom the reader has become acquainted on the above pages. When Norman heard about my book, he, just for our friendship's sake, volunteered to help me with the English version. I translated the original Finnish text into some sort of "English", and Norman then rewrote the whole story in real English. Nobody could have done it better. He understood amazingly well what I wanted to say, and did a truly Herculean job saying it in English. I am very grateful to Norman and feel most fortunate to have a friend like him.

Preface

Although scientific achievements are extensively covered in the media, scientists themselves rarely take the trouble to describe scientific progress from their own point of view. James Watson is a notable exception, for in *The Double Helix* he gives the story behind one of the greatest discoveries of this century, the elucidation of the structure and function of the gene. His book also tells of the intense competition between various research workers and groups during the years in which the secrets of "the spiral thread of life" were finally unravelled. In addition to the exceptional importance of the achievement, the book has been popular because of the rare opportunity it gives to look into the world of creative science.

The topic of my own research is interferon, a small protein with very remarkable properties. It inhibits the multiplication of viruses, regulates cell division, and plays a central role in the defence mechanisms in our body. A lot is now known about it, but many of its secrets still await discovery. If unravelling the structure of the gene is compared to a sprint, then the interferon story resembles a marathon run, but not one that has merely involved miles of monotonous jogging, for in it there have been moments of triumph as well as despair in the lives of many scientists.

My goal in this book is then to tell about my own life with interferon. I hope it will give the reader another glimpse into the world of science and the life of the scientist, into the way research is carried out in the laboratory and the clinic, and how progress in science comes from international collaboration. I have also tried to show how the mind of a scientist operates and how he experiences success and failure. The book is largely based on the 37 files containing my correspondence from 1962 to the present day. It tells about the pursuit of new basic knowledge, but the main theme of the story is the application of new knowledge for the benefit of medicine.

This book is addressed to a reader who is interested in the world of science, but he or she should not need to know even the basics of biology or medicine to follow the story. If some parts are found difficult to follow, the reader can without hesitation skip over them without losing the thread of the story.

Interferon's journey from the laboratory to the pharmacy shelf, from basic research to its acceptance as an antiviral and anticancer drug, took three decades. This book tells how the combined efforts of scientists from all around the world work towards this common goal.

Contents

1

The Way Through An Impasse

On Thursday, February 17th, 1972 a small elderly man sat in my office in National Public Health Institute in Helsinki, Finland, smoking a pipe. He had had a long journey and was suffering from jet lag. Nevertheless we had a lively conversation, for we had a common interest and a common problem. Samuel Graff had recently retired from the Chair of Biochemistry at Columbia University, New York. He had remained keen to continue with research work, and thanks to his colleague, David Habif, Professor of Surgery at Columbia, an opportunity turned up.

Habif was a brilliant surgeon who had operated on countless cancer patients during his career. All too often he had seen a distressing final outcome to his technically perfect surgery; his patients had succumbed to cancer cells which had spread to sites beyond the reach of his knife. This problem led him increasingly to consider whether the body's natural defence mechanisms could be harnessed in some way so as to prevent such recurrences. In particular, he became an enthusiast for interferon, then apparently an almost magical substance. It seemed to have so much potential for use in medicine, but none was available in amounts or even in a form that made it possible to test it in patients. Habif had a wealthy friend, Seymour Milstein, and had been able to persuade the latter to provide the finances needed to support an interferon laboratory for Graff. In this laboratory, Sam had grown cancer cells and tried to make interferon from them, but he had run into many difficulties.

In my laboratory, we had been struggling with the production of interferon for nine years, learning little by little how best to produce this exotic substance. We had succeeded in making it in much larger amounts than elsewhere, but we still had not found how to purify it. I had pinned my hopes on collaboration with chemists in England and the United States, but we had all run into apparently unsurmountable obstacles, and seemed to have reached a dead end.

Sam and I talked all that day and the next. At one point during our discussions, he took down a volume with a brown cover from my bookcase, found the page he was looking for, and said: "Perhaps this is the way forward."

I was quite familiar with this book, which contained the papers delivered at a conference organized by The New York Academy of Sciences in 1969. I also knew the article which Sam pointed out: it was written by Karl Fantes, a scientist in London, with whom I had collaborated for several years. The point to which Sam drew my attention was Karl's sentence: "Interferon activity can be precipitated with ammonium sulphate, methanol, ethanol or acetone, but without worthwhile purification." I had read the sentence many times before, but not seen anything particularly interesting in it. Sam's idea that ethanol, the chemical name for ordinary alcohol, could be used to purify interferon did not strike me as very important.

Sam flew back to New York where sadly, his future was dogged by many misfortunes. I returned to my busy daily routine, and his comment about ethanol was soon lost in the ocean of other things in my mind. Yet, for some strange reason it kept coming back into my consciousness over and over again. The very next week I started to try using ethanol to purify our interferon. The results surpassed my wildest expectations. A way was opened through the impasse. In a short time we had worked out an effective purification method. In the following decade, nearly all clinical studies with interferon carried out in all parts of the world used material made and purified in Finland.

When Sam visited me, interferon had already been under study for 15 years in many countries, and it was another 15 years before it became registered as a drug for use against cancer and viral diseases. If Sam had not drawn my attention to ethanol in February 1972, I believe this period would have been even longer.

2

Chance

Chance has played a big part in my life. At high school, I was such an enthusiastic gymnast that at one time, I even contemplated a career teaching this subject. However, an accident during one summer vacation put paid to these plans: I broke my left arm badly and the fracture did not unite perfectly. This ended my ambitions as a gymnast, but fortunately, I was left with little permanent disability.

I then decided to become a physician, influenced by my best friend who decided on the same career. For me, the main attraction was the prestige and general high regard accorded to members of the medical profession. Perhaps also at the back of my mind, thoughts of the beautiful lake-side house and tennis court of our family doctor in Mäntyharju, the place where I was born, played a part in my decision. At the time, my intention was of course to engage in clinical practice: I knew nothing about a career in medical science, and indeed could not even imagine what this might involve.

During my medical studies in 1953, we had to perform a series of laboratory experiments as part of the bacteriology course. My particular group of students was supervised by an assistant physician, by name Mauri Parmala. The head of the department, Professor Olli Renkonen, had made it the custom to select students at random to give short talks on specific topics.

At that time, there was no separate Chair of Virology at the University of Helsinki. The subject was included in the serology-bacteriology course, and taught by two excellent lecturers, Kari Penttinen and Nils Oker-Blom. I particularly liked Penttinen's informal "American" style of presentation.

Penttinen knew he was by no means an expert on the subject of bacterial viruses, and he decided to follow his Professor's example and make one of us prepare a lecture on this topic. As he did not know us very well, he asked Parmala, who had conducted the practical course, to advise which student should be chosen to talk. Mauri looked round, and as I happened to be standing nearby said: "Why not ask that chap, Cantell." That sentence had far-reaching consequences on the subsequent course of my life.

I cannot now remember if I was pleased or upset at having been selected. However, as I had no choice, I set to work to prepare my lecture, working in the evenings in the departmental library. The topic began to fascinate me and I spent many hours reading. I still vividly recall one particular article, written by the French scientist, André Lwoff, and published in a journal, *Endeavour*. In it, Lwoff described how bacterial viruses could become integrated into the genetic material of the host bacteria and function as if part of it, but could nevertheless be reactivated under certain circumstances to become again independently multiplying viruses. This paper made a big impression on me. Might human cells similarly harbour integrated viruses? Could these play an important role also in human diseases?

I was not surprised when Lwoff was later awarded a Nobel prize. *Endeavour* also retained a special glamour in my eyes. So I was particularly pleased when, nearly a quarter of a century later, I was invited to contribute to this journal, even though it then no longer had quite the same cachet as during its previous golden age.

My talk on bacterial viruses went moderately well. My classmates probably gained very little from it, but it had one important result: Penttinen invited me to spend some time working in his laboratory. I was attracted by his personality, and the little I had learned about viruses had given me an interest in research. We arranged to meet in his laboratory, and so my career started.

The reader will repeatedly come across the part that chance has played at various times in my career, and indeed, the importance of chance as a factor in scientific progress should not be belittled. Penicillin, one of the most useful drugs known to man, was discovered as the result of a chance accident, and chance also played a role in the discovery of interferon. Small chance events have a big role in determining health and disease. Even the origin of life on this earth and its subsequent development may perhaps be merely the result of a series of chance events.

3

Into The World Of Laboratory Life

Kari Penttinen spent 1948 and 1949 studying virology in the laboratory of Thomas Francis in Ann Arbor, Michigan. When he returned to the State Serum Institute in Helsinki, he first worked on influenza viruses, but soon also on mumps virus. Some years previously, George Hirst, working in New York, had discovered an interesting property of these viruses: they were able to agglutinate, i.e., to clump together red blood cells. This phenomenon made it easy to reveal the presence of virus to the naked eye, even though the virus particles themselves were far too small to be seen even under a microscope. This agglutination effect could also be used to measure amounts of virus quantitatively, and it became the subject of my first laboratory experiments.

Although my medical studies took most of my time, I began to spend the evenings and weekends in the small virus laboratory at the State Serum Institute. We used red cells obtained from chicken blood in routine agglutination experiments, and previous studies had shown that cells from different birds might give different results. My first research problem was to find out the basis for these differences between individual chickens and to show their practical importance. It led to my first scientific paper, published in the Annales Medicinae Experimentalis et Biologiae Fenniae in 1955, with Kari Penttinen, Veikko Tommila and Kari

Cantell as its authors. Viewed from today, the paper has only rather slight scientific value, but at the time, it meant a great deal to me. I had performed the experiments with my own hands, and I had slowly and laboriously prepared the first draft of the manuscript. Finally it was ready, and I waited on tenterhooks for Penttinen's opinion. His verdict that only a few manage to write so well during their whole lifetime, was a pronouncement that undoubtedly was meant mainly to encourage a young and ambitious beginner in research. In this, it certainly succeeded, and it gave me confidence that I might one day be able to carry out worthwhile research. Nevertheless, my sights were still set on becoming a medical practitioner.

At this time, virologists everywhere began increasingly to make use of cells grown in culture for their studies. In 1949, John Enders and his co-workers had shown that polio virus multiplied in cultures of monkey kidney cells and destroyed them. Their observations had far-reaching consequences, and earned them richly deserved Nobel prizes. They made it possible for the first time to make polio vaccines, but they also showed the way in which many other viruses might be grown in the laboratory.

Penttinen also started cell culture work in his laboratory, and he told me about the new plaque method developed in Renato Dulbecco's laboratory in the U.S. When a sheet of cultured cells is infected under suitable conditions with a virus suspension, individual virus particles form colonies, which can be made visible to the naked eye as patches or "plaques"; by counting the number of plaques formed, the number of infectious virus particles in the suspension can be precisely measured. Another benefit of the technique is that the progeny formed by an individual virus particle can be isolated. Later, Dulbecco was also awarded a Nobel prize.

Realizing the importance of Dulbecco's work, Penttinen asked me to try to set up the technique. It proved to be far too difficult, for at the time we lacked the equipment, the skill and the experience. Fortunately, I soon appreciated the situation, and so did not waste too much time on problems which I could not then overcome. Later in the United States, I mastered the technique and I have since used it extensively for more than 30 years.

In the mid-1950s, a Norwegian scientist, Oscar Lahelle, described how to culture cells detached from the amnion, the inner membrane of the human placenta. These amnion cell cultures proved to be a good substrate for the growth of many viruses, and Penttinen felt it would be very useful to have them in his laboratory. He therefore sent me to Oslo to learn the method.

When I came back, I obtained a placenta from the Obstetrics Department and processed it exactly as I had been taught. The first experiment was a great success, and everyone in the laboratory came to look at the beautiful cells in my culture bottles and tubes. Unfortunately, this initial success was followed by numerous failures. All in all, I cannot take much pride from my early achievements in the demanding field of cell cultures.

As well as my medical studies and my research activities, I developed a keen interest in philosophy. I attended the lectures of the famous Finnish philosopher, Eino Kaila, I read books on philosophy and I made friends with many philosophy students. For a while, I wondered if I had chosen the right field: should I move from medicine to philosophy?.

4

I Become Involved With Mumps Virus

Mumps is one of the oldest known diseases of man. The writings ascribed to the father of the medicine, Hippocrates, which date back to the fifth century B.C contain a description of a typical case in a man which was complicated by a mumps virus infection of the testicles (mumps orchitis); certainly, this disease would have plagued man long before that era.

Mumps used to be a considerable problem in the Finnish Defence Forces. Much of this country is sparsely populated and often, children living in such areas are not exposed to mumps during childhood. Consequently, many young men started their period of compulsory military service without having developed a natural immunity. The result was that mumps infections were rife among the recruits, with around one thousand cases each year. About one patient in every five developed orchitis, a painful and long-lasting complication which was not infrequently followed by hormonal disturbances, and even by sterility.

Worrying over this problem, Kari Penttinen came to the conclusion that the only way to stop the epidemics would be to vaccinate the recruits. No mumps vaccine was then commercially available, and so he decided to make some himself. It was already known that mumps virus could be cultivated in fertilised hens' eggs, and Penttinen used this method to make a stock of virus, which he rendered non-infectious by treatment with formalin. In 1953, his "home-made" vaccine was tested in a small-scale pilot study in army recruits. The results were so encouraging that he decided to make enough mumps vaccine to immunise all military personnel.

At that time, plans were being drawn up for a new building for the State Serum Institute, in which it would become feasible to produce the vaccine on a large scale. Accordingly, and in spite of my inexperience, Penttinen decided to give me the task of determining what factors influence the growth of mumps virus in eggs. The goal was to work out the best conditions to use when making the vaccine in the future.

I was given a free rein with this task. I worked on my own, and all too rarely sought help from Penttinen. For I am by nature something of a lone wolf, stubborn and independent, and then, as now, one who prefers to work on his own. Indeed, the strengths and weaknesses of much of my later research probably reflect these characteristics. However, the immediate result was that I wasted much time on pointless experiments, and made many stupid mistakes as the result of sheer ignorance. One of these, nevertheless, led me to discover a strange and interesting phenomenon. I will explain the background.

Unlike a bacterium, a virus particle can multiply only within a living cell, and so some suitable cells must be available if the virus is to be grown in the laboratory. A technical advance in the late 1930s showed that many viruses could grow in the cells of chicken embryonic membranes. When a fertilised hen's egg is incubated at body temperature, two of these membranes form sacs surrounding the developing embryo, the amnion, and outside this, the allantois, and both contain a small quantity of fluid. It was found that the cells in these embryonic membranes could be infected by injecting a virus suspension through a small hole drilled in the egg shell. The hole is sealed with paraffin wax, and the eggs are re-incubated at 37°C. The virus infects the cells in the embryonic membranes, and the newly formed virus particles are released into the allantoic or amniotic fluid and can be collected.

Mumps virus is routinely grown in this way in the allantoic cells of 7-day-old chick embryos. Five days after infection, the allantoic fluid in a single egg contains an enormous quantity of virus, even as much as 10^{11} (one hundred thousand million) virus particles, or theoretically enough to infect the whole population of the world.

The mistake I made was to take allantoic fluid containing mumps virus, collected from one lot of infected eggs, and inject this without any dilution into a further set of eggs, and to continue to pass the virus as undiluted allantoic fluid into further eggs in series in the same way. I reasoned that the more virus particles I injected into my eggs, the greater the certainty of infecting the maximum number of cells, and so of getting a very big yield of progeny viruses.

Not being then aware of the relevant scientific literature, I did not know what had been found in the very same situation with influenza viruses. Here, experience had shown that, unexpectedly, the biggest yields of virus came from eggs infected with very few virus particles, and in practice with a high dilution, usually one part in ten thousand, of a virus-containing allantoic fluid.

When in my mumps studies I injected undiluted allantoic fluid into the eggs, there were at first no problems, and the virus yields were quite good. However, after 5 or 6 successive transfers of undiluted allantoic fluid, the yields dropped sharply. I nevertheless blindly went on making further serial passages with undiluted inocula for reasons I no longer remember. To my surprise, the virus yields started to increase and again reached high levels.

There were differences of up to 100 000 times between the highest and lowest virus yields in the harvests from successive transfers of allantoic fluid, and the fluctuation in virus followed a predictable pattern. Why was virus multiplication sometimes impaired? An experiment which I performed threw some light on the problem, but again as I write this, some 35 years later, I cannot recall exactly what led me to carry it out.

I used a particular allantoic fluid mumps virus preparation and showed that it gave only a poor yield of virus when injected undiluted into eggs. But when I diluted the same virus preparation ten thousand times before inoculating it, the yield was excellent. This showed that the virus inoculum must contain some factor capable of inhibiting the growth of mumps virus. What could this inhibitory factor be?

A number of experiments convinced me that virus particles that had been inactivated and effectively "killed", were not responsible for blocking the multiplication of the infectious virus particles. Furthermore, the inhibitory factor still had some effect even when the allantoic fluid inoculum was diluted until it contained fewer virus particles than the number of cells in the allantoic membranes. Thus the inhibitory factor did not appear to be directly associated with the virus particles themselves.

In my routine work, I allowed the mumps virus to grow in the chick embryonic membrane cells for 5 days. To my surprise, when the time of incubation was shortened to only 3 or 2 days, the quantity of the inhibitory factor was greater. So this mysterious factor on which I had stumbled also seemed unlikely to be something derived from degraded virus particles. What then could it be?

5

Interferon Comes On The Scene

$\mathbf{A}$t this time, I knew very little about virology. I therefore tried to rectify this by reading as much as possible, and so spent much time in the library of the Department of Sero-bacteriology on the second floor of 24, Fabian street. The library was small but comfortable, and I enjoyed the deep leather chairs, the subdued light, and the smell of tobacco.

During the final stages of my medical studies, I was made an assistant demonstrator in the Department, and from early 1957, shortly after I graduated, I worked in this building as the first Assistant in the newly established Department of Virology. There were two applicants for the Chair in this new Department, my teacher, Kari Penttinen, and another lecturer in Virology, Nils Oker-Blom. The selection committee took their time in making their choice. Meanwhile, as an Assistant, I had few routine duties and enjoyed a regular salary.

I spent long days, and often weekends as well in the laboratory. My girl friend, Aila, frequently came too when she had finished her work as a hospital nurse to see what I was doing and to keep me company. From her point of view, we certainly could have found more pleasant things to do together. We became engaged on August 31st 1957, and Aila adapted herself bravely to her role as the fiancee of a young scientist.

One day in the library in the autumn of 1957, I came across an article in the English journal, *The Proceedings of the Royal Society* which greatly excited me. It was by Alick Isaacs and Jean Lindenmann from the Medical Research Council's laboratories in London, and it had the title "Virus interference I. The interferon". I still vividly remember what a big impression this article made on me. I realised at once that this important study was highly relevant to my own research studies. I read the paper again and again, but even then did not understand its full significance, which indeed I only appreciated many years later. What then was this discovery?

Isaacs and Lindenmann had studied what had been a subject for research since 1935, the phenomenon of virus interference. This is the name given when one virus is seen to block the growth of another virus when both try to infect the same cells. The original observations were made in experimental animals infected simultaneously with two different viruses. Under these circumstances, sometimes only one of the viruses produced its characteristic disease, and the infection with the other had no effects.

Most studies on the interference phenomenon had made use of influenza viruses, and leading the research in this field were two scientists, Werner and Gertrude Henle, who worked in Philadelphia. I shall say more later about this husband and wife team, for they played an important part in both my personal and scientific life. In 1943, they published a scientific article which had far-reaching consequences: they showed that a virus can produce the interference phenomenon even when it has itself been rendered incapable of growth; the virus can be inactivated by any of a variety of treatments and yet still retain its capacity to interfere with another virus. They showed this by injecting the inactivated virus into embryonated eggs, i.e., eggs containing a fertilised embryo; it did not grow but could still block the multiplication of another virus.

The use of inactivated viruses made it much easier to study the interference phenomenon, and during the 1940s, the Henles themselves made a very detailed study of the factors which affect it. Their work culminated in a key review by Werner Henle "Interference phenomena between animal viruses" published in 1950.

In their experiments, Isaacs and Lindenmann used small pieces of chick allantoic membrane to provide the living cells needed for the growth of influenza virus and also carried out interference studies with these. They used this culture method mainly for reasons of economy: each little bit of allantois, incubated in an individual test tube, substituted for the use of a whole embryonated egg, and many pieces could be prepared from embryonic membrane of a single egg — both research funds and eggs were still in short supply in post-war England.

In the U.S.A., hens' eggs and research funds were not in such a short supply, and so the Henles used eggs in their interference studies. This point had a crucial impact on their results, for if they similarly had used embryonic membrane cultures, they would undoubtedly have discovered interferon, and probably some ten years sooner.

In their study, Isaacs and Lindenmann planned to follow how the interfering activity of an inactivated influenza virus preparation decreased as it bound to the allantoic cells in their cultures and penetrated into them. To their great surprise, there was no such decrease: the interfering activity in the culture fluids actually increased as incubation was continued, exactly the opposite result of what they had expected. They soon showed that this increase in interfering activity had nothing to do with the virus particles themselves, but resulted from some unknown factor which was released from the cells and accumulated in the nutrient medium surrounding their allantoic pieces. They called the factor "interferon", because of its apparent role in the interference phenomenon.

Isaacs and his colleagues soon established some of the basic properties of their interferon. It appeared to be a protein; it could inhibit the multiplication of many different viruses; it had no direct effect on a virus outside a cell, but instead produced changes within the cell such that the subsequent growth of a virus was blocked. The cells themselves seemed to suffer no deleterious effects in consequence.

It is certain that long before Isaacs and Lindenmann made their discovery, many other scientists including the Henles must have produced interferon during the course of their experiments, but they did not recognise its presence because their assay methods were too insensitive or because they wrongly interpreted their experimental data.

6

My Thesis — And Other Matters!

The original goal of my own studies had been to optimise the growth of mumps virus in eggs. From this point of view, my efforts certainly provided useful information and gave me experience which proved very helpful when later I actually prepared the vaccine. However, my scientific interest rapidly moved from the original topic to the virus-inhibiting factor which I had come across accidentally. What was the relation of my "factor" to interferon? I began to compile data for an academic thesis.

In the spring of 1958, I had joined some 20 other participants from various countries in Europe on a one month "Training Course in Virology" arranged by the World Health Organisation. The first half of course was held in the laboratories of the Central Public Health Laboratory in Colindale, London, and directed by a distinguished elderly virologist, Fred MacCallum, whom I greatly admired. I knew that as long ago as 1937, he had published a paper on viral interference.

Most of those on the course were not very proficient in English. I soon made friends with two of the participants, an Italian, Adalberto Felici, and a Pole, Jerzy Georgiades and together, we had some enjoyable times sightseeing in London.

While in London, I heard that Isaacs, the discoverer of interferon, was going to give a lecture at a meeting of the Royal

Society. I went to this and waited with interest to see what sort of person Isaacs would prove to be. In the publications, I had noted the first name was Alick, and in my ignorance, I assumed that this, like Alice, was a female name. So I could hardly believe my eyes when the person who mounted the rostrum to give the talk turned out to be a small, slight, dark-haired man.

During the second two weeks of the course, the participants went to various laboratories in different parts of England to learn practical work at the bench. I discovered that there would be a place for one of us in Isaacs' laboratory in Mill Hill, and I greatly hoped to be the lucky one. But when I spoke to Professor MacCallum, I was greatly disappointed to hear that my quick-witted and voluble friend, Adalberto, had already secured this place. Instead, I went to Liverpool, and I remember only that my short stay in that grey northern city was rather dreary.

Aila and I were married just before midsummer in Aila's home town Oulu, in Northern Finland. Our honeymoon was limited to the sleeper train journey from Oulu to Helsinki, for I was also wedded to my work, and striving frantically to finish my doctoral thesis before my period of compulsory military service. I had postponed this to the age of 26 years; the unavoidable deadline was the autumn of 1958.

The prospect of eleven months of military service fitted very awkwardly into my personal time table, but it never occurred to me to try to get out of this obligation. For I had spent my boyhood during the war with Russia, and the headquarters of the Finnish army was located in my own home town Mikkeli. The heavy bombardments forced my family to move to my birthplace, Mäntyharju. I remember vividly the coffins that brought back home from the Eastern front so many of the young men from Mäntyharju, and the emotional atmosphere at these soldiers' funerals. Towards the end of the war, the ominous and uninterrupted thunder of guns was carried from the front. My mind was filled with anxiety and worry. What was going to happen to us? What would become of my country? I have strong feelings of patriotism which stem from those dark times. I have spent my life working in a very international field, but I have never found a conflict between being a patriot and a part of an international community.

I sent my thesis off to the printer just before I started my military service. One day in March 1959, I had to defend my academic dissertation formally in public. I was at that time attending the training school for non-commissioned officers, and Corporal Kari Cantell defended his thesis in auditorium XII of the main building of the University of Helsinki, wearing the uniform of the Häme cavalry regiment. My fellow-soldiers viewed my affairs with no little astonishment: "This chap is given a grant and spends it on buying 10 000 eggs!"

The whole of that week was eventful. I began my leave by visiting Aila in the Obstetrics hospital, for she was suffering from toxemia of pregnancy. Just as I arrived, she went into labour, and so I was able to be present during the arrival of our first child, a healthy boy. The vivid experience of joy at his birth is something that has remained with me ever since.

At the end of the week, Aila was able to attend the thesis ceremonies at the university, but sadly, she was still too weak to take her place at the celebration dinner: the toxemia had taken its toll.

7

Should I Become A Scientist?

After my military service, I obtained a position as an Assistant Physician in the Department of Virology at the State Serum Institute, where my chief was Kari Penttinen. The new Institute had just been finished, and so it was time to start to realise his plan to make mumps vaccine. This was the task for which I would be paid. The work went ahead smoothly, and a programme for the general vaccination of all military personnel against mumps started in 1960. It continued for a quarter of a century until a combined measles, mumps and rubella (MMR) vaccine became available, and finally replaced our domestically produced mumps vaccine.

The mumps vaccination virtually eliminated mumps and its much-dreaded complications from army personnel. There was a resultant economic benefit which was easy to calculate, but of course it was impossible to assess how much human suffering was avoided. Who can measure the cash value of a testicle? Overall, Penttinen's vaccine was a great success. It was a bold and original achievement, and a fine chapter in the history of Finnish medicine. Later, Penttinen was awarded the Matti Äyräpää prize, the highest recognition for medical research in Finland.

Apart from my routine responsibilities, I was keen to get back again into interferon research. My aim was to try to produce interferon in the cell cultures now in routine use in our laboratory. For this purpose, I decided to use HeLa cells, a line of cells which originated from a cancer of the uterus in an American woman, who had long since died from its consequences. Her cancer cells live to this day in culture in laboratories all around the world.

Alick Isaacs had recently put forward the hypothesis that cancer cells in general were neither capable of producing interferon nor could be protected by interferon against virus infections. However, I soon found that HeLa cells did indeed produce interferon, and also were protected by it against viruses. My results were accepted for publication in an internationally read virology journal.

The possibility that interferon might be used as an antiviral drug attracted investigators from the very beginning. Virus infections were very common, and vaccines are available only to protect against a few. Thus there is a great need for effective antiviral drugs. However, none were then even on the horizon, apart from interferon, which looked promising in many respects. Laboratory studies had shown that this inhibited the growth not only of influenza viruses but also of almost all other known viruses, while at the same time it appeared to have no toxic effects on cells. So big hopes were pinned on it as a potential antiviral drug. In England, three pharmaceutical companies joined in a collaboration with the British Medical Research Council to try to turn the Isaacs and Lindenmann discovery into something which could be used in medicine. An additional spur was given to the work by the knowledge that another great British discovery, penicillin, had in the end been developed into a commercial product in the United States.

I was myself greatly intrigued by the possible eventual uses of interferon in medicine. I came across a new scientific paper in which Isaacs described how he had injected interferon into the skin of rabbits and thus prevented the growth of the smallpox vaccine virus, vaccinia, when this was later injected into the same site. This finding had no practical relevance, but it did show that interferon could act in the body i.e., *in vivo*, and not only under the artificial conditions of cell culture (*in vitro*).

The practical problem at that time was that no worthwhile amounts of interferon were obtainable. Interferon is formed only in living cells, each cell only producing very little and for a short time. So to prepare a substantial amount of interferon, huge numbers of cells would be needed. Furthermore in 1959, David Tyrrell, a co-worker of Isaacs, had shown that interferon works best in the cells of the species from which it is derived: interferon for human use would have to be prepared in human cells.

At that time, it was quite impossible to visualise how enough interferon could be obtained to protect an experimental animal, let alone a human being, against viruses. I wondered whether there would be any practical value in trying to protect just a small local site. This made me think of the eye, and the occasional infections of the superficial cells, the cornea, by such viruses as vaccinia, herpes and adenovirus. Vaccinia virus was then widely used for smallpox vaccination, and was sometimes accidentally transmitted from the vaccination site to the eye. Much more frequent and so more important as a cause of eye disease was herpes simplex virus. After a first infection, this virus frequently remains in the body throughout life, normally in a latent form, giving no signs of its presence. However from time to time, and for reasons still not fully understood, the dormant virus is reactivated, and then produces blisters in the skin — often around the mouth or lips ("cold sores") or in the genital area. It can also, though less commonly, produce a painful ulcer on the surface of the cornea. Such an ulcer usually heals within two weeks, even in the absence of treatment, but may leave a permanent scar on the cornea; as the infection may recur many times, there may be progressive impairment of vision, and finally blindness. This type of herpetic eye disease is not uncommon, especially in Southern countries.

Could interferon drops put into the eye prevent the development of a virus infection? I decided to experiment. I prepared rabbit interferon by treating cultures of rabbit kidney cells with mumps virus. In terms of today, its content of interferon was only minuscule. I persuaded a colleague, Veikko Tommila, to collaborate with me. He had also written a thesis on a virological subject under the tutelage of Kari Penttinen, but had then decided to specialise in ophthalmology. We infected the eyes of four rabbits with herpes virus and of another four with vaccinia virus. Then we instilled drops of my rabbit interferon six times into the right eye of each rabbit and "mock interferon", a control preparation with no interferon content, into the left eye. These treatments were repeated daily for four days. Every other day, the rabbits were taken to the Department of Ophthalmology where Veikko had the proper equipment for examining the eyes of our "patients". To our disappointment, interferon therapy had no effect on the course of the herpes infection. However, we had some consolation, as in all rabbits, the vaccinia infection was clearly less severe in the interferon-treated eye than in the control eye. Interferon had given some partial protection against the disease caused by that virus.

At that time, there was much criticism of the quality of experimental medical research in Finland, and an example was given of a publication which had five authors, involved two rabbits, and reported negative results. Perhaps our own work was dangerously near the bottom of the barrel, but nevertheless, the partially positive results obtained by two workers with four rabbits were accepted for publication in the prestigious English medical journal, *The Lancet*.

We planned further studies along the same lines, but for various reasons given below, they were not pursued. Nevertheless, the idea of using interferon to treat eye diseases stayed in the back of my mind.

During these studies my attitude towards research began to change. I became increasingly interested in interferon and my experiments were going well. My doctoral thesis had been well received by referees in Finland. To my surprise, I got a letter from the United States inviting me to write a review on mumps virus for the important series "Advances in Virus Research". I was convinced there must have been some misunderstanding and I wrote to the editor to voice my suspicion, but he reconfirmed his invitation. I took this as an honour and a challenge, and so took the work very seriously. After the thesis and my military service, this new task took up almost all my free time, and left me with very little for my young wife and our tiny son.

From all points of view, I became more and more attracted to research, and I began to consider a career as a scientist. Some small encouragements and recognitions which I received were a further influence. Although my post as an Assistant Physician at the State Serum Institute was perhaps not ideal, it did give me the chance to carry out some research work in addition to my routine duties. The prospects for promotion in the field of virology did not look very promising, but neither Aila nor I found that too much of a concern.

8

To The New World

$\mathbf{A}$s I began to lean more and more towards a career in science, I realised this would inevitably mean a long spell studying in a foreign country. For at the beginning of the 1960s, virology in Finland was still almost in its infancy, and it was essential for a young virologist to train and gain further experience abroad.

Where should I go? There were some excellent virus laboratories in Europe, but for various reasons, I was convinced I should go to the United States. This seemed more glamorous, and also, because I would be eligible for an ASLA fellowship, it would probably be easier to obtain support for a stay across the Atlantic than for a spell in a laboratory in Europe. Even with our small child, Aila was quite prepared for the adventure.

When it came to deciding on the laboratory where I should try to seek a position, the choice seemed quite easy. While himself studying in the United States, my present instructor, Kari Penttinen, had met Herbert Morgan. The latter was working with mumps viruses and had discovered that these viruses cannot only clump red blood cells together, like influenza and some other related viruses, but also and uniquely cause the red cells to lyse, i.e., to disintegrate, leaking their content of the red pigment, haemoglobin. Morgan told Penttinen that I would be welcome to work in his laboratory. After the initial excitement, I began to have some doubts about this prospect. My interests had moved from mumps viruses to focus on the topic of interferon, and I decided to look for myself into the possibility of getting a position in an American laboratory interested in this. There were not many choices and, in my opinion, the Philadelphia laboratory of the husband and wife team of Werner and Gertrude Henle was by far the best.

While I was attending an international congress in Copenhagen, I had the chance to meet Werner, and with shaking knees, I introduced myself and told him how much I would like to work in his laboratory. To my great delight, his response was friendly and positive: indeed, he immediately promised me a position, with only the timing and the financial arrangements remaining to be settled. I wrote to Morgan to say that, after all, I would not be coming to his laboratory, which not surprisingly, did not please him.

While at this same congress, I spoke with Alick Isaacs. He had just published a paper which implied that interferon could not protect cancer cells against a virus infection. I told him about my experiments with HeLa cells. The results contradicted his new hypothesis, and the information can hardly have been very welcome, but in spite of this, he was friendly and interested in my data.

The practical problems relating to our trans-Atlantic journey were easily solved. I passed the language examination and was awarded an ASLA fellowship. Aila made arrangements to get a U.S. work permit, and through the international nursing exchange program, obtained a post in the Woman's Medical College of Pennsylvania. We took with us a young Finnish girl, Pirjo Halonen, to take care of Heikki, and in August 1960, started our journey on the cheapest flight available, a propeller plane of the Icelandic company, Loftleidir. The flight was very long and interrupted by successive landings in Oslo, Reykjavik and Newfoundland.

There was much turbulence, and I suffered miserably from air sickness. Aila proved tough and robust and took good care of Heikki who, fortunately, was very little upset by the journey. After 36 hours we finally arrived, exhausted, in New York and passed through the customs and immigration formalities. There, we had a most unwelcome surprise. My good friend, Tapani Vainio, was then working in Philadelphia, and had planned to meet us at the airport and drive us to Philadelphia. He was nowhere to be seen. Finally, a phone call to Philadelphia established that he had just come back from the airport in New York after a misunderstanding about the arrival time for our flight.

At long, long last, we reached the end of the journey. We stayed with the Vainios for a while until we found a place of our own. This was in Germantown, a district which white people were leaving while African American people were moving in. We bought a second-hand silver-gray Chevrolet Corvair, which filled me with pride. A new, exciting and important phase in our life had begun.

9

An Embarrassing Episode

I still vividly remember an incident in 1960, before we left for the United States. The year had begun well: both my period of military service and my thesis were behind me; we had a healthy son; I had obtained a permanent appointment which gave me scope to undertake research; and my U.S. travel plans were going well. Among all this, something very embarrassing happened.

I was invited to give a talk on interferon at a meeting of the Finnish Society for Pathology, to be held in the Institute of Anatomy of Helsinki University. This flourishing society counted among its members almost all Finnish pathologists and microbiologists, and many attended its regular meetings which were followed by a meal at some nearby restaurant.

As the time for the meeting drew near, I was very busy in the laboratory. At the same time, our domestic routine was disrupted because Heikki had had a reaction after his routine smallpox vaccination, and in consequence, I spent several rather restless nights. When at last the day came, and I was waiting for my turn to speak, I began to realise I had not made adequate preparations. Instead of writing my talk out or practising it, I had merely sketched an outline, with the intention of speaking freely from my rough notes. I felt my blood pressure begin to rise. I became increasingly nervous, my mouth was dry and I broke out in a cold sweat. When my turn came, I managed to get started, but I could not keep my thoughts clear, and remember what to say next. I started to panic and feel dizzy, and finally I had to abandon my presentation. I was utterly mortified and ashamed of myself. How could this have happened? How could I think of becoming a professional scientist if I was incapable of giving a lecture in public. I was deeply upset by the fiasco, and my self-confidence suffered a severe dent. Indeed, in a way it was a relief to leave Finland and move to the New World where I felt I could start again.

Looking back now, I can see that I learnt an important lesson from this wretched episode. It taught me the importance of preparing properly for any public occasion. Since that time, it has been my custom to speak without notes, but I always spend a great deal of time in advance in planning, writing and practising my talks. Later on, I was awarded high marks for my official test lectures for the positions of Docent and University Professor, and my lectures were rated highly by the medical students. Indeed, over the years, I believe that I have acquired a reputation as a good speaker. If this is correct, this was not achieved easily, but over the years and little by little, these positive indicators have gradually restored my self-confidence.

I remember talking with Kalle Österlund, a well known Finnish doctor who for many years ran the medical television program "Home doctor". In the course of this, he interviewed hundreds of physicians. He told me that in his experience, interviews with shy and sensitive people usually turned out well, because they came well prepared for what they considered something of an ordeal. Problems were much more likely with those who were very self-assured, for they made no advance preparations, and often, their ex tempore remarks led them into difficulties.

10

The Henles' Laboratory In Philadelphia

Werner Henle and his wife, Gertrude, were born in Germany, and both studied medicine in Heidelberg. Werner came from a Jewish family. His grandfather, Jacob Henle was a famous scientist who lives in the history of physiology and medicine as the discoverer of what is still termed the loop of Henle in the kidney. Werner's father was an eminent professor of surgery, whose patients included many celebrities from Germany and far afield, and he was a contemporary and colleague of Ferdinand Sauerbruch.

As Hitler's power increased in the 1930s, Werner was forced to leave Germany. With his wife, he moved first briefly to Cuba, then ruled by the dictator, Fulgencio Batista, and from there to the United States, where they both obtained positions in the research department of the Children's Hospital of Philadelphia. This became the venue for their life's work. During the second world war and for many years after, they worked in an improvised laboratory in the cellar of the old Children's Hospital building, and there they carried out pioneering studies on influenza and mumps viruses and on viral interference.

The Henles were dedicated to their work. They had no children, whether involuntarily or from design I never discovered, and we,

31

their students, became their substitute. Many of them came from Germany, a country which the Henles seemed to regard with rather mixed feelings. Werner was a natural aristocrat of old times, a master photographer, and an expert in music and literature. I admired him greatly.

The Children's Hospital at that time was located in an area in down-town Philadelphia, a depressing slum district inhabited by poor people, mostly African American. Violent crimes were commonplace, and very few scientists ventured to return to the laboratories after dark. As a naive Finn, I found much to astonish me. At first, I did not take the warnings I was given very seriously, but I soon realised how different the environment was from that back at home. One rainy day, I saw a black boy in the very act of stealing my car and driving it away. I was furious, but also in despair as I could not possibly afford to buy a replacement. To my surprise and joy, I later got the car back undamaged.

The laboratories were small and modest, but the work carried out in them was of the highest calibre. I was given a place with one of the staff, Kurt Paucker, and I worked with Kurt for nearly two years. There were just the two of us and a single technician, Sandy, later replaced by Elsa, in a single room. Its door was adorned with a metal plate with the label "Benjamin Rush laboratory". I never discovered what sort of a man Dr. Rush had been, but "nomen est omen" did not seem to apply in this instance. We worked diligently enough, but there was certainly no sense of hurry.

11

My Jewish Friend, Kurt

Reflecting again after all this time on the two years I spent each working day with Kurt Paucker in the intimacy of our small laboratory, I realise even more clearly what different background we had.

Kurt was born of wealthy Jewish parents in 1924, during the Weimar republic, and grew up in Berlin, where his grandfather, Ignatz, had founded a successful business. In the 1930s, the rise of the National Socialists began to cast a shadow on the family. When Kurt was 8-years old, his older brother, Arnold, persuaded him to sew an anti-fascist movement badge on his swimming trunks. The result of this was that he was severely beaten up by Nazi thugs. Finally, in 1933, Arnold and Kurt were banned from attending school because they were Jews. It was clear there was no future for the boys in Germany, and in 1936, Arnold was sent to Palestine. Two years later Kurt, then aged 14, left for France, where he attended a school for Jews near Paris. When France capitulated to the Nazis in 1940, most of the pupils were sent sooner or later to work in salt mines in Germany, or to the gas chambers in Auschwitz. Kurt and another boy were the only ones in the whole school who managed to get away. They joined the resistance movement, and Kurt was given the task of escorting a group of women and children to Switzerland. Once there, he got in touch with relatives in Zurich, and as a result was able to go to school and university in that country.

Many of Kurt's relatives in Germany were killed; his parents escaped at the last moment to Shanghai, but it was only ten years later that they were able to join Kurt. During these years, he had emigrated to the United States, where he first worked in a bank as a translator. He managed to save enough money from his salary to cover the costs of university studies, and he married Ellie, a Jewess from New York. In 1952, he joined Werner Henle as a postgraduate student, working for the degree of Doctor of Philosophy, and he was awarded his Ph.D. degree three years later.

I first came across Kurt at the International Congress of Microbiology held in Stockholm in 1958. I did not talk to him then, but I listened to his presentation, and my first impression was far from favourable. Indeed, I found his manner very self-assertive and irritating, and frankly, I did not take to him at all. Later, I came to realise how wrong my first impression had been.

Kurt was a man without roots, a Jew and very much a citizen of the world. My own background was in total contrast. I been part of a small circle of friends and relatives in a small and remote country on the periphery of the western world, and my perspectives were narrow in many ways. Kurt was the first Jew I ever got to know and conversely, for him, I was the first Finn. Was it even remotely likely that a close relationship would develop between two such different people? In fact, in spite of all the differences, our collaboration worked excellently, and very soon we became fast friends. We worked together closely in the laboratory during the day, and often our families met during our free time.

I came to admire Kurt, and I learnt much from him. He was a brilliant organiser, who made detailed plans for the next day and so on into the next week. When I arrived in Philadelphia, I was something of a workaholic, but Kurt soon showed me that there are other precious things in life, and that one can do good research without needing to spend all one's evenings and weekends in the laboratory.

Kurt was a man of many talents. He was extremely good with his hands, and I remember vividly how skilfully he packaged things that needed to be sent overseas. He had mastered several languages, and he had many interests. For my own part, I detest window shopping and browsing through stores. As a result, I usually buy, for example, the first pair of shoes which happen to fit. However, Kurt, like Aila, was quite the opposite, and the two of them thoroughly enjoyed shopping expeditions, and especially visits to art shops.

Kurt was an exceptionally honest and forthright person with strong views about many topics. These views he expressed freely and without regard to what other people might think. For example, he was a savage critic of some facets of the social conditions in the United States at that time, and this often involved him in arguments with those staff in Henle's laboratory who were native-born U.S. citizens. On the other hand, he strongly emphasised the many good things in the U.S.

Kurt had been denied many opportunities during his life, and he was openly bitter about this. "All good things in life come too late," was a phrase he was fond of repeating. But this darker side of his personality was offset by warm humanity, genuine friendliness and a delightful sense of humour.

Kurt had a deep hatred for Germany, and my own similar feeling about Soviet Russia helped me to understand his attitude. I believe that in the end, but only after many heated discussions, I finally managed to get him to understand why, in the final stages of the Second World War, my own country had been an ally of the Germany which he so hated.

Through Kurt I learned much about Jews and Judaism, and I began to appreciate the splendid contributions made by Jews to the arts and sciences. I have since become close friends with many Jews. Altogether, my stay in the United States and my friendship with Kurt have meant a great deal to me both as a scientist and as a human being.

The VIII grade of Mikkeli lyceum in fall 1949. I was the smallest in my class during all the eight years at that school. I graduated at the age of 17, and during the following years I grew in height over at least half of my former class mates.

Engaged in August 31, 1957: Aila Kokko and Kari Cantell.

In the beginning of the common road. Our wedding was in the cathedral of Oulu on June 18, 1958.

My teacher of virology: Kari Penttinen.

A corporal in the cathedra. The official defense of my doctoral thesis in March 1959.

The discoverers of interferon: Alick Isaacs and Jean Lindenmann.

Our family before the journey to the United States in 1960.

With Kurt Paucker in Czechoslovakia in 1964. An international congress of interferon was held in the castle of Smolenice in the background.

Gertrude and Werner Henle. I worked in their laboratory in Philadelphia in 1960–62. The Henles had to leave their home country Germany in the 1930s, because Werner was a Jew. In the U.S. the couple made important contributions to virology.

A talented scientist and a dear friend of mine, Tapani Vainio whose career broke in the midst of its rise in summer 1965.

Charles Chany

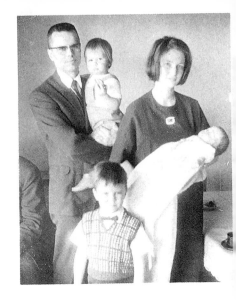

The director of State Serum Institute and my boss over a decade, an edged strongman, Professor Eero Uroma.

My growing family in 1964.

Ion Gresser, one of the great pioneers of interferon research, showed in 1961 that white blood cells are able to produce interferon.

Participants of the Smolenice congress in 1964. From front left: Kurt Paucker, Kari Cantell and Martti Valle. The second row: Ed Kilbourne, Jaqueline De Maeyer-Guignard and Edward De Maeyer.

Wetterkulla Medical Center cooling after sauna in spring 1965. For more than thirty years this group of scientists has gathered twice a year in Wetterkulla mansion in Eräjärvi, Finland, to write and discuss. It has become an important group of friends in my life. The reader of the book will get acquainted with many of the members of the enter. From left: Osmo Järvi, Erkki Saxén, Bo Thorell, Esko Nikkilä, Harri Nevanlinna, Sulo Toivonen, Kari Cantell, Tapani Vainio, Kari Cantell, Tapani Vainio, Jussi Saukkonen, Kari Penttinen. Lauri Saxén who took the picture is missing from it.

Tapani Vainio, Kari Cantell and Jussi Saukkonen planning a symposium on nucleic acids in Wetterkulla in fall 1964 (Photo: Lauri Saxén).

Hans Strander from Sweden worked for five years in my laboratory in the 1960s. With him we developed the methodology for the production of white cell interferon, and he started to test the Finnish interferon in patients in Stockholm. I have shared many joys and sorrows with Hans.

H.R. Nevanlinna — the father of the Finnish Blood Service — in the 1960s. The preparation of interferon in Finland is based on his life's work.

Maurice R. Hilleman whose sensational finding changed the course of interferon research in 1967.

Thomas C. Merigan, a bright scientist in the laboratory and the clinic, carried important clinical studies with Finnish interferon in the U.S. and in England in the 1970s.

At the meeting of Lake Placid in 1973. From left:
Paul Came, Karl Fantes, Kari Cantell (moustache
phase) and Hans Strander (before quitting
smoking).

Annikka Hirvonen — my skilful co-worker
over 30 years.

Foreign co-workers in my laboratory in the
beginning of the 1970s. Front: Erik Mogensen,
back: Francis Ekiko from Cameroun.

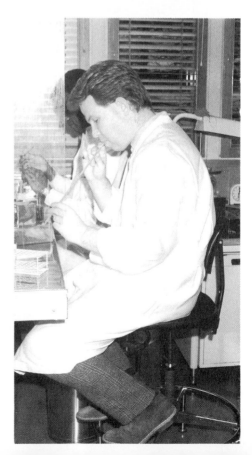

With Mathilde Krim in April 1975. My suitcase was lost in London airport; therefore I am in a sweater in a reception at the home of Mathilde and Arthus Krim. (Photo: Paul Schumach).

Professor Tsunataro Kishida and his wife Kyoko from Kyoto are friends of Finland.

Professor Yasuiti Nagano, the grand old man of interferon research in Japan, and the author in Kyoto in 1975.

The best yields of interferon are obtained in round bottles. (Photo: Terence Spencer).

Sendai virus is cultivated in embryonated eggs. (Photo: Terence Spencer).

Hanna-Leena Kauppinen has directed the production of interferon in the Finnish Red Cross Blood Transfusion Service for over 20 years. Our collaboration has been excellent. The picture is from 1974.

The cover page of the first issue of Tiede 2000 — journal in 1980.

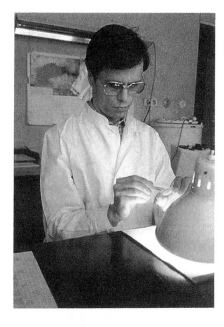

Counting virus plaques. I have done a lot of this work in the course of 30 years.

Professor Charles Weissmann transferred interferon genes from a Finnish blood donor into bacteria. In his laboratory in Zurich the coli bacteria began to produce human interferon on Christmas 1979.

With Aila, John Hiscott and Charles Weissmann's godson Michel Streuli in 1984.

Interferon meeting of WHO in Geneva in March 1982. From left: Norman Finter, Alfons Billiau and Kari Cantell.

Jordan Gutterman tested Finnish interferon for the treatment of leukemia in Houston, Texas. The breakthrough studies of his group led to the official acceptance of interferon as an anticancer drug.

With our friends Elizabeth and David Habif and Seymour Milstein in Helsinki in 1976. David Habif, professor of surgery at Columbia University, studied the use of interferon for the treatment of breast cancer during 20 years until his death. The Milstein Award which is given annually to young interferon scientists bears the name of Seymour Milstein. (Photo: Vivian Milstein).

Englishman Geoff Scott with his wife and Aila on a street of Rome in 1985. Geoff Scott has instilled interferon into many noses in the course of years.

Co-chairing a meeting with Charles Chany in Kobe, Japan in 1983. The speaker is M. Canivet from Paris.

German Rainer Sundmacher clarified in ten years how herpes infections of the eye can be treated with interferon drops.

Fidel Castro, interpreter, Manuel Limonta and Kari Cantell at the inauguration of the Interferon Institute in Havanna.

Bill Stewart with his daughter Vivian and Aila and Kari Cantell in 1986. Bill has had both friends and foes in the international interferon community.

Pondering interferon problems with Harri Nevanlinna in Wetterkulla in spring 1979. (Photo: Lauri Saxén).

In lighter spirit with Harri Nevanlinna and Erkki Saxén (left) in Wetterkulla in fall 1980. (Photo: Lauri Saxén).

The happy grandparents in spring 1985. (Photo: Lauri Saxén).

Scientific discussion with my friend Lauri Saxén in 1991. (Photo: Erkki Saxén).

Aila and the grandchildren Joonas, Julia and Aino in 1992.

12

Working With Kurt

While a student with the Henles, and later as one of their co-workers, Kurt had for a number of years studied the virus interference phenomenon, working with preparations of an inactivated influenza virus and fertilised hen's eggs. I timed my arrival in his laboratory very well, for when I joined him, he was starting to switch to carrying out experiments in mammalian cell cultures. For about a year, Kurt had worked with a Polish fellow, Zofia Skurska, growing L cells, a particular line of mouse cells, in suspension cultures. In such cultures, the cells are constantly but gently agitated, and this prevents them from attaching to the glass or plastic of the container. Soon, they adapt to growing freely and multiplying in suspension in the nutrient culture medium. This has the great advantage that the cells can be kept growing at a steady state, as is easily checked by taking samples at frequent intervals and counting the number of cells. Kurt and I made plans to carry out a detailed quantitative analysis of the phenomenon of viral interference in this experimental system.

Zofia was near the end of her fellowship and due soon to return to Poland, but her work with Kurt had been fruitful and the system had begun to work well. It proved easy to induce the interference phenomenon in the mouse L cells by treating them with an inactivated preparation of Newcastle disease virus (this belongs to the same general group as the mumps and influenza viruses). Kurt had recently acquired a sophisticated and at that time newly developed instrument called a Coulter counter which made it possible to count the number of cells in a sample almost instantaneously, and to assess the size of the cells. Within a few years, such electronic cell counters became standard items even in small routine laboratories, but in 1960, a Coulter counter was an exotic novelty, and Kurt was very proud to own one.

Kurt and I approached science in a rather different way. Kurt was very dexterous and had a flair for laboratory work. He planned experiments very systematically, but proceeded rather slowly, one step at a time. My ideas were more adventurous, but only a very few proved fruitful. Nevertheless, although Kurt was older and had more laboratory experience than I, we collaborated from the very beginning as equals, and we made a good team, each complementing and developing the skills of the other. Kurt had not been broken by the many hard blows in his life. He had a healthy self-confidence and a strong, well balanced character. I was the more ambitious and more prepared to take risks. We had a mutual respect for one another, and this led to a happy atmosphere in our laboratory.

From my own point of view, it was a stroke of luck that I was able to benefit from the experimental system that others had put in place. Almost at once, we had success, making a number of novel observations relating to interference and interferon. The fruits of our two years' work together was six publications in the leading scientific journal, *Virology*.

We noticed that cells did not yield a "second crop" of interferon, even when they were retreated with a virus capable of giving a good yield with fresh cells. Jan Vilcek, a Slovak scientist then working on interferon in Bratislava but who later escaped to the United States, made the same observation independently.

We also noted that cultures, which had been treated with interferon, similarly lost their capacity to produce interferon, but after two cell divisions, the granddaughter cells were again able to make it. It had previously been suggested that interference is an "all or none" phenomenon — in other words, that a cell is either fully protected against viruses or fully susceptible. We were able to show convincingly that this was not so. We studied the descendants of cells which had been completely protected by interferon, and looked at the successive generations to see if they were still protected. When challenged with a virus, the cells first regained their ability to produce virus proteins, but only after seven cell divisions were the cells again completely susceptible and able to form fully infectious virus; after six divisions, a cell had given rise to 64 descendants which were still partially protected against viruses.

We made our most important discovery soon after my arrival in Philadelphia. This was not the result of a sudden inspiration, but rather something we stumbled upon because of our particular experimental system. We noticed that cells treated with interferon, and also those producing interferon, increased in number and divided at a rate that was obviously slower than that of controlled untreated L cells. A single dose of interferon had a minor and transient effect, but if cells were grown in the continued presence of interferon, the effect was dramatic, and eventually they completely stopped dividing. Nevertheless, if interferon was then removed from the culture, the cells slowly regained their normal growth rate.

Although the results were very clear and reproducible, we took great pains to prove that the phenomenon was really due to interferon rather than some impurity in the preparations which we then had available. We could have proven our point conclusively if we had been able to repeat our experiments using pure mouse interferon. Unfortunately, this only became available some 10–15 years later, but when at last it could be tested this way, our original finding was finally confirmed. In the meantime, and as we had anticipated, our results were criticised or even ignored, which we found irritating, and, indeed, unreasonable. One main reason was that at that time, many scientists regarded interferons as highly specific antiviral substances which had no detrimental effects whatsoever on cells, and they were very reluctant to give this concept up.

Our paper was published in 1962, with Kurt Paucker, Kari Cantell and Werner Henle as its authors. Some 25 years later, this paper was listed as a Citation Classic, and as the only surviving author, I was invited to write a brief account of the background and origin of the study. I wrote this in a sad frame of mind, remembering with affection my dear co-workers and friends, Kurt and Werner.

I must mention another paper that Kurt and I co-authored. Alick Isaacs and his co-workers in London, and also Jean Lindenmann, who had returned to Switzerland, had tried to immunise experimental animals with interferon. When they tested blood from these animals, they found no signs that any had formed anti-interferon antibodies, i.e., had specifically modified blood proteins that reacted with interferon and neutralised its effects. Both these groups drew the conclusion that interferons must be poor stimulators of immunity, or perhaps were completely non-antigenic, and incapable of eliciting the formation of antibodies.

The interferon made in our mouse cell cultures was quite easy to concentrate ten times. We used most of what we made for our cell culture experiments, but gradually we accumulated some surplus. We began to inject this surplus material into three guinea pigs, and from time to time, we bled the animals and tested the samples for interferon antibodies. Although we had no great expectations, to our great delight and surprise, we found anti-interferon antibodies in samples of blood obtained after five months of immunisation from all three guinea pigs: this was the first convincing demonstration that interferon can indeed evoke the formation of antibodies.

We also noticed that antibody detection was crucially dependent on details of the assay method, and that if too much interferon was used in the tests, small amounts of antibody escaped detection. It is probable that the earlier failures to produce antibodies resulted from injecting interferon in insufficient amounts and over too short a period. Nevertheless, assay insensitivity may well have been an additional factor that contributed to the apparent absence of antibodies.

Antibodies have become very important tools in interferon research. By using them, interferons can be classified, purified and assayed, and the role played by interferons in different diseases can be elucidated. On the other hand, in a few patients being treated with an interferon preparation, its benefits are blocked due to the development of antibodies, and this can be a troublesome clinical problem.

There is one report in the literature which may have been an earlier description of the formation of antibodies to an interferon. This came from two Japanese scientists, Nagano and Kojima, who, like several other research groups (as a review of the older scientific literature shows clearly), described some of the effects of an interferon before Isaacs and Lindenmann published their paper in 1957. The Japanese workers found that the tissues of rabbits infected with vaccinia virus contained an activity that inhibited the multiplication of vaccinia viruses, and they called this activity "facteur inhibiteur." Their report was published in French in a French journal which is not widely read, and for this reason was largely overlooked. Facteur inhibiteur may indeed well have been an interferon, but it was not characterised sufficiently well for this to be certain. I believe it would have attracted much more attention if it had been shown active against viruses other than vaccinia viruses: this would have convincingly ruled out any possible immunological basis for the activity against vaccinia virus, and shown that facteur inhibiteur is a completely new kind of virus inhibitor. In Japan, Professor Yasuiti Nagano has received recognition and credit for the first discovery of interferon, but elsewhere this is almost universally attributed to Isaacs and Lindenmann, as Nagano himself has noted with disappointment.

In one of their early papers, Nagano and Kojima reported that their facteur inhibiteur, prepared in rabbits, led to the formation of antibodies when injected into other rabbits. On the basis of present knowledge, an interferon is a very poor inducer of antibodies in animals of the same species as that from which it is derived, and I am personally still unconvinced that they actually obtained antibodies against an interferon.

13

Great And Small Prophets

While I was working with the Henles, I met many scientists from all over the world who came to visit their laboratory. I came across many others when I attended seminars at the nearby Wistar Institute, where my friend, Tapani Vainio, and later another fellow countryman, Eero Saksela, worked. Together with Kurt, I also participated at important scientific meetings, one of the Cold Spring Harbor Symposium series and a meeting of the Federated U.S. Microbiological Societies. It was an experience for me to see in person so many of those whose scientific papers I had read. I had often visualised these scientists in my mind, but reality and my mental picture seldom agreed well.

The first of the visitors to the Henles whom I remember was Charles Chany from Paris. He had read my paper dealing with interferon and HeLa cells and was scathing in his criticism. He suggested that I had not measured effects of interferon at all, but instead inhibition of virus growth caused by undue acidity in the culture medium, a trap, which he said he had himself fallen into. I was deeply astonished and hurt, and was quite unable to convince him that my results were genuine. Indeed, his criticism stung me so much that I sent to Finland for the HeLa cells which I had used in my study and with Kurt showed that these, together with another line of HeLa cells from a U.S. laboratory, were as

sensitive as normal human cells to the action of interferon. Our
paper helped to demolish finally the hypothesis about interferon
and cancer cells which Isaacs had propounded.

I nursed resentment of Chany for a long time, and in turn
adopted a suspicious view of his studies. However, a quarter of
a century later, during an interferon meeting in Washington, the
two of us dined alone together. Charles relaxed and told me about
his life, and his experiences as a Hungarian Jew in the Buchenwald
concentration camp. The result of the evening was that at long
last I completely shed my hard feelings towards him.

In contrast, Jean Lindenmann's visit left pleasant memories.
He had dinner with us at home, and we got on to familiar terms.
To my regret, he left working in the field of interferon soon
afterwards, though he came back to it again many years later.

I met Alick Isaacs only a few times at various scientific meetings,
and I did not have a chance to get well acquainted with him.
It is my opinion that the collaboration between Alick Isaacs and
his Swiss research fellow, Jean Lindenmann, was fundamental to
their discovery of interferon, and would not have been made
by either working on his own.

My friend at the Wistar Institute, Tapani Vainio, became
increasingly interested in interferon and transmitted this to his
boss, the new and dynamic director of the Institute, Hilary
Koprowski. They studied the role of interferon in the resistance
of mice against virus infections, and obtained intriguing results.
This was a source of a special delight to me, since I looked forward
to future collaboration with Vainio back home in Finland.

One day the Henle laboratory was visited by Frank Burnet,
a scientist from Australia. Then already world-famous, Burnet
became even more famous later on when he was awarded a Nobel
prize, and became Sir MacFarlane. I had read his books and
papers, and I knew that Isaacs had studied in his laboratory. He
came with Werner Henle to our laboratory, and Kurt and I

described our studies to the guest. Henle then asked Burnet what future he foresaw for interferon as a drug. Burnet replied "None at all," and Werner agreed with him. They both regarded interferon as something extremely interesting from the theoretical point of view, but did not believe that it would find any clinical applications. I decided not to open my mouth, but in spite of my great respect for these two eminent scientists I silently strongly disagreed with them. The idea of trying to use interferon to prevent or treat viral diseases had seldom been far from my thoughts ever since my studies in the eyes of rabbits. I realised, of course, there would be tremendous difficulties, but this was the goal I was determined to pursue.

14

Back Home Again

I had originally planned to spend just a year in Philadelphia, but Aila and I found our life there most congenial, and in the laboratory, my studies were making good progress. Werner Henle thought so too. He arranged financial support to enable me to spend a second year with him, and would have extended it even further. My second year proved particularly fruitful as far as research was concerned, and Aila, who had picked up English amazingly quickly, was getting on very well at her work. We started to contemplate spending a much longer time in the United States, preferably in Philadelphia, but some surprising news from Finland settled the question.

The director of the State Serum Institute in Helsinki was Professor Eero Uroma. He was a most unusual man with a rugged character that is not easy to summarise in a few words. His greatest achievement was to force through, in the face of great obstacles, the construction of the new and magnificent State Serum Institute in Helsinki. The main opponent of the scheme was a personal enemy, Olli Renkonen, Professor of Sero-bacteriology at the University of Helsinki, who regarded the new State Institute as potentially a dangerous competitor with his own department. In this at least, his judgement proved correct! Renkonen brought all his prestige and political connections into play to try to prevent or delay the development of the Institute. In the final stages in the debate, the opponents of the scheme tried to force through a proposal to site the new institute in Jyväskylä, a town in middle Finland and a considerable distance from Helsinki. As I understand it, Renkonen's guiding principle was not the national benefit but rather the struggle for power and his personal antipathy for Uroma. To be fair, however, it must be said that as one who had greatly developed his own institute and who had inspired many young scientists, he had great merits.

In his turn, Uroma was a tough and ruthless opponent of Renkonen. As the Director of his own institute, he was an autocrat — indeed almost a dictator — who often failed to distinguish the importance of big and small issues. He spent much time talking with politicians in the cafeteria of the parliament building, and in these circles he had the reputation of being somewhat peculiar, but a brilliant scientist. He also had a private clinical practice for patients suffering from skin complaints and sexually transmitted diseases which took up much of his time. In his own Institute, he was regarded as a politician rather than a scientist, but he valued research work highly.

There are many anecdotes that cast light on his colourful personality. A visiting politician asked him what sort of people worked in his institute. "Damned good people," he replied "the cream of the country, first class as individuals but unanimous in one respect: they all criticise me." The last phrase was indeed completely true.

In one of his books, the well-known Finnish writer, Mika Waltari, has described the young Uroma, or Uggeli as he was called, when they were together during their military service in 1930: "Uggeli's capacity to talk passes belief. Not even the major can keep him in check. In fact, he goes on talking until we all fall asleep."

I had found myself in a rather strange position; both as a student and subsequently, I had worked both in the University department and at the State Serum Institute. Being young and a starry-eyed idealist, I found it sad that there was so much animosity between the directors of the two institutes, which was in turn mirrored by the members of their staff.

After I had carried out my public defence of my doctoral thesis, there was a formal official banquet, and the two longstanding enemies, Renkonen and Uroma, were both present, though they did not exchange a single word. In my after-dinner speech I expressed the pious hope that the relations between the institutes would improve, as I had personally experienced great warmth and friendliness in both.

It was a considerable surprise to me and many others, when Kari Penttinen came second to Nils Oker-Blom in the competition for the Professorship of Virology. The applicants seemed very equally matched, but I gained the impression that Renkonen had had his fingers in the pie, and that Penttinen had suffered from the fact that he worked at Uroma's institute. Oker-Blom had a brilliant university career, starting in the bacteriological laboratory of the Helsinki city hospital, and rising to the most senior post in any Finnish university, the Chancellorship of the University of Helsinki. At the climax to his career, the almost unique title of "archiater" was bestowed on him — there was only one contemporary physician with this title in all Finland. Nisse Oker-Blom was an excellent leader in the department of virology, and his great personal qualities, as well as those of Penttinen, are shown by the fact that the two former competitors later worked in close harmony as professors in the same institute.

During the time when I was in Philadelphia, Penttinen, who was a specialist in internal medicine, was the head of the department of Virology at the State Serum Institute. In this position, like the heads of many other departments, he found himself more or less continuously at loggerheads with Uroma. In 1962, I got the totally unexpected news that he was going to resign his position in order to become the senior physician in a large insurance company Salama for which he had previously worked part time. He had become so thoroughly tired of the continuous strife with Uroma that when there was a vacancy for a permanent position with Salama, he made the radical decision to move there. It was fortunate for virology in Finland that he was later made a Professor in the University.

Penttinen's departure must have been a blow to Uroma, and he did not have much choice in appointing a successor. There were indeed very few virologists in the whole country, and two of these already worked in his own institute, Kaisa Lapinleimu and myself. Lapinleimu had written her doctoral thesis on the subject of polio viruses. She was older than I, but her specialisation in pediatrics and her responsibilities as mother of a large family had occupied much of her time, and as a result this energetic lady had only a short list of scientific publications to her credit. Furthermore, in personality and sex, Lapinleimu made little attraction for the chauvinistic Uroma when he looked for a new department head. So his thoughts turned towards me. Meanwhile, my friend Tapani Vainio had returned home. He wrote to Philadelphia to explain the situation and urged me to apply for the post. It seemed an opportunity that I could not afford to miss. I sent in an application, and at the age of 29 years, I was appointed head of the Department of Virology at the State Serum Institute. Three years later, the post was upgraded to professorial rank.

I found the situation somewhat confusing. On the one hand, I was obviously delighted to have a good and permanent position, with several laboratories and many staff under my control. On the other, I felt very young and inexperienced in my position as a director, with responsibility for administration, virus diagnostic services, epidemiological studies and problems relating to vaccines.

It was also not very agreeable to be subordinate to Uroma. Deep down in my heart, I still longed for a post at Helsinki university.

Later, a second Chair in Virology in Finland was established in Turku, the old capital city. I applied for the post, and was considered well-qualified for it, but another applicant, Pekka Halonen, was appointed. Frankly, he was the best man for the post. Later, I twice came close to appointment as a professor at the University of Helsinki or as a research professor of the Finnish Academy, but I finally had no success, and a real research post remained a dream. So I gave up further attempts, reconciled myself to the situation and remained a government employee in the Institute, which has twice since changed its name, becoming first, the Central Public Health Laboratory and later, the National Public Health Institute. When I was a young man, I never dreamt that I should spend my whole working life up to the day of my retirement in the same institute, but that is how it has turned out.

15

Chick Interferon

When I moved back from Philadelphia to the State Serum Institute, my days as a full time scientist were over, but I tried, nevertheless, to devote as much time and energy to research on interferon as my other duties allowed.

While I was in the United States, it had become quite clear to me that the scanty supplies of available interferon were a major bottleneck to research. Furthermore, all work in the field had been carried out with grossly impure preparations with very low biological activity. No one had till then seriously tackled the formidable task of making interferon in larger amounts and of higher purity. This seemed an avenue of work that was open for me.

My first decision was to decide which interferon I should try to prepare, and this proved easy: chick interferon seemed for many reasons to be the natural choice. In my department, large numbers of fertilised hens' eggs were used in the production of mumps vaccine and for influenza studies, and some of these could readily be used to make interferon. Also, I had some personal experience in culturing viruses in eggs. I set to work immediately, growing many different viruses in turn in the allantoic sac, and measuring how much interferon appeared in the allantoic fluid.

A medical student, Martti Valle, soon joined the work, and he became a permanent friend and collaborator. Martti prepared cultures of chick embryo cells, and established a good method for measuring the amount of a virus, the plaque method, which could readily be adapted to assay amounts of interferon. The plaque method formed the basis of his doctoral thesis.

Out of all the viruses we tried, we found a particular influenza virus, the strain called influenza B Lee, to be the best inducer of interferon in eggs. We improved and standardised the production method until we obtained good yields consistently, and then we were ready to prepare chick interferon in bulk. First, however, additional laboratory space and money were needed, as well as some new equipment. Fortunately, the Institute director, Professor Uroma, came to share our enthusiasm, and at once all the practical problems could be solved. Extra laboratory space was found on the top floor of the building, new technicians were engaged, and enormous egg incubators were purchased. The next problem was to get these last into the new laboratory. Uroma obtained the services of a group of sturdy firemen, who dragged the incubators to the fifth floor, leaving gouges on the walls of the stairwell all the way up which are still there today.

We began to accumulate large amounts of chick interferon, using tens of thousands of eggs for the purpose. Production ran quite smoothly, and over one hundred batches were prepared in the course of two years. Uroma harnessed two biochemists from the Institute's staff, Jussi Saukkonen and Rafik Schakir, to work on the purification of the product, using the methods developed by Hilleman's group in the American pharmaceutical company, Merck, Sharpe and Dohme.

Soon we began to appreciate what a difficult task we had set ourselves. One problem was that at that time, we had no idea how far we were from obtaining a pure product. On the basis of today's knowledge, our best preparations probably had an interferon content of only the order of 0.1 per cent, which may seem a very modest achievement. In fact, it was a good beginning, and no one else had achieved anything better. However, to the increasing discomfort of the interferon team, it became obvious that Uroma wanted to take personal charge of the project, even though his expertise, commitment and available time did not match the requirements of the difficult task. For my part, I decided to start work on the production of interferon from human cells.

16

Human Interferon

It would be wrong for me to ascribe my switch from studying chick interferon to human interferon entirely to Uroma's intervention. In my eyes, the latter was by far the most interesting to study of all the interferons, and the one that was needed in bulk. From the medical point of view, it would be important to try to find out what roles human interferon plays in health and disease. Also, only human interferon would be suitable for trying to treat patients with various diseases. Furthermore, an interferon is quite specific in terms of the animal species in which it is active, and this must reflect structural differences between, for example, the interferons derived from human cells and from chicken cells. Thus even when the structure of chick interferon had been worked out in detail, this would probably give only a general guide to that of human interferon, which would still have to be determined. For all these reasons, I began to experiment with the production of interferon in cultures of various human cells.

My first results were not encouraging. All the cells did produce some interferon, but the amounts were depressingly small. I therefore tested other human cells derived from placentas and fetal tissues, and various "continuous" cell lines, i.e., cells of the type which can be subcultured indefinitely. I also tried various technical tricks, such as culturing the cells in different ways, and stimulating them with many different viruses, in the hope that these might increase the interferon yield. Nothing led to any improvement in the depressingly low yields.

On May 8, 1963 my luck changed. I carried out an experiment which led to a new era in my life: for the first time, I used white cells obtained from fresh human blood for the production of interferon.

Already in 1961, Ion Gresser, working in the United States, had reported that these, like most other human cells, could be induced to make interferon, and in his paper he speculated on the possible role that the interferon made by white cells might play in the defence of the body against virus infections. He also suggested that the production of interferon by white cells might be of value for diagnosing a virus infection. I found his observations had the greatest interest from another point of view. Might these cells be good producers of interferon?

There was a small department in our Institute in which human blood proteins, such as albumin and gamma globulin, were purified, under the direction of a biochemist, Eva Vahvaselkä. On May 8th, Eva gently centrifuged a bottle of freshly donated blood for me. This forced the blood cells to the bottom of the bottle. She removed the liquid plasma from above the layer of sedimented red cells, and this contained many white cells, as well as a few red blood cells. She centrifuged the plasma again, forcing all the cells to the bottom of a tube which she gave to me, together with some of the red cells from the original centrifuged blood bottle, which still contained many white cells. I diluted the cells in a nutrient medium and distributed them into several tubes. I added a virus, influenza or Sendai, to each tube, and then incubated them all overnight at 37°C, mounted on a revolving drum, which kept the cells in constant movement and prevented them from settling on the walls of the tubes.

Half of the experiment failed right away. The cells cultured from the plasma turned into a gelatinous mass, probably because of the numerous blood platelets present in this cell fraction. My lab notes succinctly describe my feelings: "Clotted — damnation." The cells harvested from the red cell layer behaved better, and I measured how much interferon they had formed. The assay results came through on May 26th and they were sensational: there was no detectable interferon in the cultures infected with influenza viruses, but the tubes treated with Sendai virus contained 1000 interferon units per millilitre, which was ten times more than the best yield I had ever obtained previously from cultured human cells. Eureka!

I realised immediately the prospects opened up by these results, but the next and crucial step was to repeat the experiment, in case some technical flaw on the first occasion accounted for the result. However, the repeat experiment gave identical results.

Quick calculations showed me that human blood is a veritable treasure house of white cells: the standard donation from one single individual contains a tremendous number of these cells, around 1000 million. To grow a comparable number of cells in the laboratory, at least 100 of the conventional large glass Roux culture bottles would be required, and this would be a laborious and expensive business. I recalled Goethe's thought in Faust. "Das Blut ist ein ganz besonderer Saft" (Blood is an absolutely remarkable fluid). How could I lay my hands on the treasure house?

Later, Ion Gresser told me that it had not occurred to him that human white cells could be used for the mass production of interferon. Alick Isaacs read Gresser's paper in 1962, and did consider this possibility, but he took the idea no further.

I decided to abandon work on chick interferon as soon as possible, although I foresaw that it would leave Uroma with a grudge against me for which I would have to pay a price. In this respect, my misgivings came true.

With friendly help from Eva Vahvaselkä, I got a steady supply of white cells for my experiments. Let me pause here for a reminiscence about "tall Eva." Her Renault and my small yellow car sat side by side all day in the Institute's parking lot. Just before Christmas, I found the following words on the ever-grimy windscreen of my car: "Merry Christmas from Eva V." The words were still there after several weeks, and Aila finally remarked that it would not be a bad idea to wash the car every now and then. But back to the white cells Eva supplied. Under what conditions would they yield the greatest amount of interferon? Clearly many, many things had to be worked out, and a long programme of work lay ahead.

Only much later did I begin to realise what good luck I had in one respect. In general, cells produce little or no interferon under normal conditions and they must be "induced," i.e., stimulated by some means, so that they are forced to produce interferon for a short while. Viruses were the first inducers of interferon to be described and are still often the best. I measured the inducing capacity of many different viruses in my white cell culture system: the best turned out to be Sendai virus, one of the two I had used in my very first experiment. Sendai is a Japanese city, and Sendai virus, originally termed hemagglutinating virus of Japan, was first isolated there from a mouse with respiratory infection. It belongs to the group of parainfluenza viruses, which includes mumps virus. Other human parainfluenza viruses produce respiratory infections, especially among children, but fortunately, Sendai virus does not cause disease in man, and so is a relatively safe virus for work in the laboratory. It was also ideal for my use as an interferon inducer, as it grows extremely well in hens' eggs, and in fact better than any other virus I know. Thus I knew it would be easy to obtain large amounts of the virus. While all this was quite obvious, I only found out much later how easily my first and decisive experiment with white cells could have failed. We had found out then that there are big differences in the interferon-inducing capacity of different strains of Sendai virus. Although there is no record of its source, the strain which we happened to have in our laboratory freezer in 1963, proved to be among the best from this point of view. If it had been one of the weakly-inducing strains, I might never have gone further with the study of interferon production in white cells.

Later, we sent at their request a sample of the virus to the American Type Culture Collection. They have called it the Cantell strain of Sendai virus, and it has been distributed under this name to interferon laboratories all around the world.

17

Tapani Vainio

Vainio's name has already made its appearance several times in this book. My colleague was two years older than I, and he was torn between the attractions of a career as a surgeon and as a scientist. He began research under the direction of Harri Nevanlinna in the State Serum Institute. He told me that he became interested in studying the serology of blood groups because he was intrigued by the complexity and obscurity of this topic. This attitude says a lot about his character: he did not shrink from difficult problems but indeed actually looked for them. As well as studying surgery and serology, this versatile individual who bubbled with energy joined the famous research team led by Professors Sulo Toivonen and Lauri Saxen. After he had presented his doctoral thesis in 1957, he gave up surgery and devoted his life to research.

Tapani, or Tapsu as he was known to friends, picked up new ideas and methods exceptionally quickly. He tackled a problem unhesitatingly and quickly gathered new data. While he was working with Toivonen and Saxen, he introduced new immunological and molecular methods which opened new approaches to their own studies. For a Finn, Tapsu was also somewhat unusual in that he found it remarkably easy to get on good terms with others. Everyone was attracted by his charm and friendliness. He always worked at such a feverish pace that it was often difficult to keep up with him. I first met him when, as a medical student, I started research in the State Serum Institute. We quickly became friends, and travelled around much in his small and decrepit Fiat car.

In 1959, the whole Vainio family, Tapsu, his wife Riitta and his son Matti, went to Philadelphia, and when my family also moved to the same city in the following year, we saw much of them. Pirjo Halonen, the Finnish girl whom we took with us, looked after the boys from both families during the days, while Aila worked in the hospital; Riitta studied modern dance, which she later introduced into Finland.

When the Vainios returned to Finland in 1961, Tapsu became the head of the Blood Group Department at the State Serum Institute, succeeding Harri Nevanlinna who had moved to the Finnish Red Cross Blood Transfusion Service. Riitta was then expecting a second child.

While in the Wistar Institute in Philadelphia, Tapsu began working with viruses, and to study whether interferon played any part in the natural resistance of mice to certain viruses. Back in Finland, he turned his attention to how the developing mouse embryo becomes resistant to virus infections.

We made plans for collaboration while we were both in Philadelphia and pursued these further when we returned to Finland. In 1963, Tapsu paid a visit to the laboratory of the famous scientist, George Klein in Stockholm, in order to learn some new techniques. While there, he talked with a medical student, Hans Strander, and Hans decided to come to work in Vainio's laboratory in Helsinki. I met him there when he had started some experiments with my interferon preparations.

My own interest in interferon was principally because it inhibited the multiplication of viruses; also, by adding viruses to cell cultures, I forced them to produce interferon, or in other words, I activated the genes that control interferon production. To Tapsu, interferon was of interest precisely because it regulated gene functions: perhaps it could be used to approach fundamental questions relating to cell differentiation. To him, interferon seemed to offer a simple and controllable way of manipulating the regulatory mechanisms of genes for experimental purposes. The idea was ingenious, and he was far ahead of his time: some 25 years later, the study of the mechanisms that control the function of the interferon genes was to become a topic at the forefront of research.

We had just reached the point at which our long-planned collaboration was about to start, when, in the evening of August 16th, 1965 the telephone rang in our house in Maunula. It was Riitta to say that Tapani had been involved in a car accident near their home in Kulosaari, and had been taken to Töölö hospital. I went there immediately, and found my co-worker, Jussi Saukkonen, already there. Together with the casualty officer on duty, Kalevi Pyörälä, we went into the intensive care unit. My friend lay on his back, naked to the waist, and with closed eyes. His chest rose and fell in time with the life-support artificial respirator, but the monitor of brain function drew a straight line. Tapani Vainio was dead.

18

Hans And Gyurka

Hans Strander had come to Helsinki to work with Vainio. Now he was faced with a decision, whether to go back to Stockholm, or to join my laboratory. When he opted for the latter alternative, he did not anticipate that he would stay for five years.

Hans had already participated in some of our studies on the production of interferon from human white cell cultures, and when he now joined me, we begun systematically analysing the factors affecting production. Some other laboratories had already looked at the formation of interferon in white cell cultures, but we had a different approach and a different goal. We saw these cells as potentially a rich source of human interferon that might be used in therapy, and so our main aim was to optimise the production of this sort of human interferon in a way that could be taken up to a very large scale. Of course, during the long course of our studies, we also followed many side tracks, but we never forgot the main target.

The news of Vainio's sudden death at the age of 35 years was a terrible shock to his numerous friends at home and overseas. As a memorial, a special issue of *Annales Medicinae Experimentalis et Biologiae Fenniae* was published in 1966, dedicated to his name. It contained articles by 87 scientists from different parts of the world. In this issue, dedicated to the memory of our friend and co-worker, Hans Strander and I published our first paper dealing with white cell interferon production, which had the title: "Production of interferon by human leukocytes in vitro." This became the most frequently cited paper of all those ever published in this journal, and later, a so-called "Citation Classic."

We continued with our systematic attempts to optimise the yields of interferon from white cell cultures, and started to accumulate useful information. Perhaps the most important new finding was that serum must be present in the cell culture medium for optimum interferon production; indeed in its absence, the white cells produced very little interferon. Chany's group in Paris made the same observation independently.

We submitted our next paper to the prestigious American publication, *Journal of Virology*, and it was rejected. The referees who reviewed the paper found no particular faults with it, but reported that our results were of no great significance or interest. We found this disappointing and indeed annoying, but for several years, we had to resign ourselves to such a negative attitude on the part of many in the scientific community. However, we remained quite sure that our work was indeed important.

To digress, at the present time, the recognition given to a scientific paper is very much dependent on the journal in which it appears. There is an unofficial ranking order for journals, which is largely determined by how often the papers published in one are cited by those who subsequently publish papers in this and other journals. In my view and experience, this ranking is often irrational, and indeed it can stultify new developments because there is a built-in bias towards whatever topics are currently much in vogue.

Thinking about this now, I realise how lucky it was for me as a young scientist that I had a permanent position. This meant that I could afford to carry on long-term research in a field not then generally appreciated. If we had to compete with other work going on in the Institute, our early studies with white cell interferon would probably not have been followed up, for their importance only became apparent after some 10 years. From my own experience, I am also highly suspicious about the application of business principles to the evaluation of scientific research. This favours short-term research that yields quick results; those who are over-prolific in writing scientific papers over those who produce fewer, but papers of more lasting value; and studies which for whatever reason are currently in vogue over the systematic exploration of some new field. It is probably difficult for a layman to appreciate how easily scientists are attracted to follow whichever are the currently fashionable trends in their field.

I have never been proficient in speaking Swedish, and became even worse during my two years in the U.S. So Hans and I conversed in the laboratory in English. However, Gyurka Hadhazy, another foreign fellow, but this time from Hungary, came to my laboratory, and with his help we tried to find out why white cells need serum in order to make interferon, and whether there are substitutes. These experiments continued for many years. When Gyurka joined us, he hardly knew any English, but spoke some German. So German became the common language in our laboratory. Hans and I were very far from fluent, but somehow we managed to communicate. Fortunately there were no native-born Germans to hear us maltreat their mother tongue.

Hans and Gyurka were very different in constitution and character. Hans was typically nordic, robust and calm, a Viking. Gyurka, in contrast, was an unusually small man with a swarthy complexion and very labile in his mood. Hans has since had a very successful career, whereas, sadly, Gyurka later had many problems. When he returned to Hungary, he remained a bachelor, and lived with his parents in Debrecen. When at last they died, his mental health began to deteriorate, and he eventually needed to go into a psychiatric hospital. Later, he died in an asylum for insane. I have been re-reading some old letters he wrote me and they are deeply moving. He referred to his stay in Finland as the happiest period of his life, and told of his love affair with a Finnish girl which did not last when he returned to Hungary. I remember him as a gentle and sensitive individual whose scientific work was characterised by originality and who had great powers of perseverance.

19

How Much Interferon Can Be Prepared In Human White Cells?

Alick Isaacs died at the beginning of 1967 at the age of 45 years, and only 10 years after he and Lindenmann had discovered interferon. I did not know him at all well, but I remember his attractive personality and he was a brilliant scientist. Even before his discovery of interferon, he had published an exceptionally clear paper dealing with the relation between virus infectious units and physical virus particles, which I had read with admiration. A year after discovery of interferon, he developed a severe depressive illness, but he recovered and in the next few hectic years, a steady stream of papers came from his laboratory. Their scientific quality varied greatly: some gave new information of fundamental value, solidly based on scientific data, while others contained speculations, sometimes based on flimsy experimental evidence. In early 1964, Isaacs suffered a hemorrhage into the brain, which was followed by episodes of profound mental disturbance; three years later, he had a further brain hemorrhage and died. At that time, many scientists had a very critical attitude toward interferon, and indeed many questioned its very existence: it was sometimes scathingly referred to as "misinterpreton" or "imaginon." All attempts till then to develop interferon into a useful drug had failed completely. I feel sure that interferon brought Isaacs moments both of triumph and of bitter disappointment. I wish it had been granted to him to see where research on interferon has now led.

In 1967, two international congresses were dedicated to his memory. The first was a Ciba Symposium in London, and among the 26 invited participants, there were two Nobel laureates, Ernst Chain, who had been closely involved with the development of penicillin, and Francis Crick, who with James Watson worked out the structure of DNA. I particularly admired the latter's brilliant conversation.

Charles Chany from Paris, who I have mentioned earlier, described how interferon could be obtained from cells from the embryonic membranes of human placentas. Otherwise little attention was paid to the topic of the mass production of interferon. I came back from this meeting with a feeling of some dissatisfaction. I could have expressed my own views in relation to several of the topics discussed, but I was not accustomed to a symposium of this type, and was too shy to say what I thought. In private, I described my own studies to a fellow participant, a well known virologist, Bill Joklik; he clearly thought them rather uninteresting.

From my own point of view, the other congress was more rewarding. It was held in June in Siena, and by invitation, I gave a talk with the title: "How much interferon can be prepared in human leukocyte suspensions?" It was later published in the Proceedings of the congress, with K. Cantell, H. Strander, Gy. Hadhazy and H.R. Nevanlinna as its authors. It summarised our four years of experience with the production of white cell interferon, and speculated on the prospects for the mass production of interferon in this way.

A fundamental point in developing our method for making interferon from human white cells was to ensure that no valuable components of the donated blood should be sacrificed in the interests of interferon production. Chany's group in Paris and Soloviev's group in Moscow were both working on a similar method at this same time, but in their procedures, the donated blood was totally given over to the production of interferon. In my view, such a method could not have any real future. It was unjustifiable to use something as precious as human blood solely for the preparation of a substance which was then of no proven medical value. Indeed, quite soon Chany's team abandoned their efforts on this topic; unfortunately, neither the work carried out later by his group on the production of interferon from human placental membranes nor that of Professor Rita in Italy ultimately proved to have any utility.

In the Soviet Union, production of interferon from white blood cells went on; a macabre element was introduced by the fact that cadaver blood was used for the purpose. Although this Soviet interferon had a very low purity and potency, it was used for a number of human experimental studies.

I was determined that when we made use of the leukocytes (i.e., white blood cells) for interferon production, the other components in blood should not be wasted. So it was my good fortune that two technical developments were made at the time in the Blood Transfusion service which made it possible to separate the white cells. Formerly, it had been the practice to collect blood in bottles, and to transfuse this whole blood to those who needed it. In the 1960s, the glass bottles were replaced by plastic bags and the concept of "component therapy" began to gain ground. This meant that, instead of whole blood, the patient received only the particular component of blood which he or she needed. This of course meant that in turn, the blood had to be fractionated, i.e., separated into its different components. To start the process, the bags of donated blood were centrifuged so that the cells, being heavier, were forced to the bottom of the bags, leaving the liquid plasma component above them. The leukocytes accumulated in a thin greyish layer called the buffy coat between these and the plasma. How could we collect these buffy coats from the centrifuged blood without disturbing the other blood components? With the help of Harri Nevanlinna and Gunnar Myllylä and the skilful technicians in the Finnish Red Cross Blood Transfusion Service, the problem was solved.

By compressing the blood bag, we were able to force the plasma through plastic tubing into another bag. This tubing was cut and the next layer, the buffy coat, was expelled into a bottle, leaving the red cells behind. Initially, we recovered in this way between 10 and 40 per cent of the total number of white cells originally present in the blood bag, but small technical modifications increased the yield to 80 to 90 per cent. The buffy coats harvested in this way contained numerous red cells which had to be removed; otherwise, they were disrupted by the Sendai virus used to induce interferon production, and the red respiratory pigment, haemoglobin, released from them contaminated the interferon. We overcame this problem by using a procedure devised by the workers in Italy: we added a chemical, ammonium chloride, which disrupted the red cells, but did not affect the capacity of the leukocytes to make interferon. In this way, we were able to obtain almost pure suspensions of white cells.

In my talk in Siena, I pointed out that more than 50 000 fresh blood bags were centrifuged annually in Finland; and that if all their white blood cells were to be used for interferon production according to our method, about 10^{10}, that is to say 10 000 million, units of interferon would, according to my calculations, be obtained. At the time, this theoretically derived figure sounded almost unbelievably huge, and I would not then have believed myself that, ten years later, much larger quantities of leukocyte interferon would actually be produced each year in Finland.

At the end of my talk, I said that large quantities of human leukocyte interferon were needed so that its chemical structure

could be worked out. Also, if it could be adequately purified,
its value for clinical purposes could be put to the test. The question
of how to make the interferon sufficiently pure worried me a
great deal; I was beginning to feel cautiously optimistic about
the production of interferon, but purification seemed then an
almost insurmountable problem.

Some of those present at the Siena meeting realised the potential
of our work. They included two workers, Norman Finter and Karl
Fantes, from England, and the rising star of American interferon
research, Tom Merigan from Stanford University in California.
Tom's fertile mind soon started to come up with ideas for initial
studies in patients, and we discussed these possibilities on several
occasions.

My family had grown. Our second son, Lauri, was born in 1962
and the third son, Aaro, in 1964. In consequence, Aila was very
much occupied in looking after the three little boys, and with
running our home in suburban Helsinki. However, by making
special arrangements, we got surrogate mothers for the children,
which enabled Aila to join me for the visit to Siena. The President
of the congress, Professor Rita, arranged a rich and interesting
programme for the spouses of participants. Aila and I also greatly
enjoyed the Tuscan landscape, the incomparable beauty of Siena
and the cultural treasures of Florence. In Siena, we were astonished
to see a painting which looked so much like one well-known to
us, called "The departure of Kullervo for war." This belongs to
the world of our national epic, Kalevala, and was painted by a
famous Finnish artist, Gallen-Kallela. Was this painting in Siena
similar by coincidence or due to plagiarism? Only years later,
I learned that when Gallen-Kallela himself visited Siena, he had
also been startled by the similarity of the painting to his, but
he had firmly refuted the possibility of plagiarism. What a
strange story.

In Siena, I met for the first time Professor Nagano, the Japanese scientist whose work I described earlier. Our conversation was greatly hampered by the fact that he spoke almost no English, while I in turn knew no French.

During the congress, the news spread like wildfire that war had broken in the Middle East. It was a great shock, especially for the Jewish participants. I recall one of them asking me with great emotion how we could all sit and occupy ourselves with such trifling matters while the very fate of Israel was at stake. The far-reaching consequence of this six day war began to become apparent even while the meeting was in progress.

20

Harri Nevanlinna

When as a young medical student, I began virus research in the old State Serum Institute building in Union street, its Blood Group Department, headed by Dr. Harri Nevanlinna, was located in the same small premises. Nevanlinna was at that time also the Director of the Finnish Red Cross Blood Transfusion Service, and later, he moved into their premises and gave up the other post.

I admired Harri from the very beginning. He came from a cultured family, well-known for mathematical talent and social achievements. He had a bright and agile mind, and at times was sharply critical of the great ones in Finnish medical establishment — he did not bow and scrape before the high and mighty. He had an endless fund of stories which he told in an incomparable way. He had a beautiful wife and a magnificent French car. From my perspective as a youngster from the provinces, he seemed to be very high above me in all respects. Nevertheless, from the very beginning, he treated me in a very friendly and helpful way; indeed, when I gave him a copy of my doctoral thesis, I wrote on the flyleaf: "To someone who is a perceptive scientist and a fair-minded man." My direct connection with him was interrupted for a number of years by my military service and my stay time in the U.S., and his move to the Blood Center, but the seeds of our friendship survived.

The Blood Transfusion service in Finland originated from the "Blood league" established by the scouts in the 1930s. At that time, whenever the need arose, one of the league would go to a hospital and donate blood for direct transfusion into a patient. During the war, the service was taken over by the Defence Forces, but they gave up control during the difficult subsequent years. Then a committee was set up to organise a national blood service and Nevanlinna served as its secretary. The committee proposed that the Finnish Red Cross should be made responsible for the supply of blood service, and in 1948, the Finnish Red Cross Blood Transfusion Service was founded, with Harri Nevanlinna as its director. There could not have been a more suitable person for the job, and Finland was fortunate that a man who measured up to its requirements was available for the demanding post.

Harri was a fine scientist, but also an energetic and brilliant organiser, with the ability to get on good terms with those in the highest positions and with ordinary people. He had a rare combination, possession of all the qualities needed to build up an enormous organisation. His insight and his clear and strong convictions that blood donation should be entirely voluntary and be supplied for transfusion use without any charge proved a firm foundation for the Service, and at the same time, his positive attitude towards research raised the international prestige of the Finnish Blood Center to a very high level. His great achievements over his 40 years of work have been widely recognised at home and abroad.

In the early 1960s, I told Harri about my interferon studies and my increasing need for white cells. I could not then give any guarantee that the Blood Center would ever benefit from my research. The early optimistic hopes for the use of interferon as a drug dating from the time of the original discovery had

long since faded away, and pharmaceutical companies had for
the most part lost interest. Most scientists were openly critical
in their attitude, or even disparaging. Both the production and
the purification of interferon looked like intractable problems.
Was it at all reasonable to expect that any significant progress
would result in this difficult field in small and remote Finland?

Harri did not hesitate for a moment, and from the very
beginning, gave my interferon studies his full support, showing
in this way his attitude towards long-term research activities. In
consequence, the Blood Center began to separate increasing
numbers of leukocytes for my studies. It meant extra work, and
the business administrators grumbled at the expense, but Harri
was as good as his word. I felt the responsibility deeply and was
full of gratitude. What a great relief it was to me when, more
than a decade later, interferon began to bring not only
money but also some international fame for the Blood Center.
Certainly, the friendship between Harri and me strengthened
over the years.

21

The Shadow Of Maurice Hilleman

Maurice R. Hilleman first became known when he isolated viruses of a previously unknown group, the adenoviruses, from recruits during outbreaks of respiratory infections in the U.S. army. Later he moved to become head of the Merck institute in West Point, Pennsylvania, where he concentrated on the development of virus vaccines. The work carried out under his direction was of considerable scientific merit and led to a number of important and commercially significant products.

Early in the 1960s, Hilleman became interested in interferon, because of its broad spectrum of antiviral action and its low toxicity. It seemed to him to have all the properties required for a successful antiviral drug, even though its species specificity posed a problem. He had under his control a team of highly competent biologists and chemists, and he set them to work on the production and purification of chick interferon. This was highly active on chick cells but devoid of any effect on human cells. It seems that Hilleman cherished the hope that, as the purification studies proceeded and material of increasing potency was obtained, the chick interferon would begin to show some activity even in human cells. This did not happen, and, furthermore, its purification turned out to be an immensely complicated task.

Hilleman made a complete turn in his attitude, and became very pessimistic about the prospects for interferon in human medicine. On the basis of experiments in cell cultures, eggs and experimental animals, he made estimates of the amount of interferon which would be needed to give even short-term protection to a human being against viruses. His calculations, which were based on many theoretical extrapolations, ended by providing figures so huge that the use of interferon as a medicine looked completely out of the question. Because he was a man with great prestige, and he wrote clearly, logically and in an easily understood style, he probably convinced most of those who read his many articles published on the subject that there was no practical future for interferon as a drug.

The generally pessimistic views about interferon at that time were strengthened by a study in human volunteers. The British Medical Research Council, together with three British pharmaceutical companies had set up a Scientific Committee on Interferon. Scientists involved in this group made interferon from cultures of rhesus monkey cells (the existence of a relationship between man and monkey is also shown by the fact that interferons from each have at least some activity in the cells of the other). Drops containing some of the monkey interferon were instilled into the noses of human volunteers, who on the following day were deliberately infected with a common cold virus. The result was clear-cut and completely negative. The interferon gave no protection whatsoever against the common cold.

It is easy to say with hindsight that the whole study should not have been done, for the potency of the interferon preparation used was far too low. Nevertheless, when the negative results were published in the prestigious medical journal, *The Lancet*, they certainly did not add to the reputation of interferon.

Nevertheless, not everyone believed Hilleman's calculations. Norman Finter in England and Ion Gresser, who had now moved to Paris, continued their persistent work, and were able to show that mice could be protected against lethal virus infections even with quite small amounts of mouse interferon. There was thus a considerable discrepancy between Hilleman's theoretical calculations and the experimental data. I believe that my presentation in Siena also helped to make attitudes towards interferon somewhat more positive. Perhaps after all, the production of vast amounts of human interferon was not entirely impossible, and perhaps very high doses of interferon were not necessarily needed to prevent virus diseases in man. There began to be again a feeling of cautious optimism about the prospects for interferon, but in 1967, startling news from Hilleman's laboratory seemed to change everything.

After a decade of interferon research, it was clear that cells produce little or no interferon under normal conditions. For interferon synthesis to begin, cells must be stimulated, "induced" according to the laboratory jargon, but even then synthesis continues only for a few hours. When it began to look as though the production and purification of interferon was going to be an extremely difficult task, many scientists began to wonder whether some inducer of interferon could be used as an alternative. It seemed attractive to think that, by giving a suitable inducer when needed, the cells in the human body could be induced to make interferon, thus obviating the laborious process of having to make interferon and then inject it. The only problem seemed to be to find a suitable inducer.

Many viruses were known to be potent inducers of interferon, but there was an obvious problem associated with their use, quite apart from any undesirable consequences they might cause in a patient: inevitably, the body would become immunised against a virus used for this purpose, and that particular one could not be used a second time as an inducer: it would be ineffective.

The sensational news from Hilleman's laboratory was that certain double-stranded nucleic acids had proved to be highly effective inducers of interferon. Such nucleic acids existed in nature, while others could be synthesised in the laboratory. One of the most potent ones was the complex between polyinosinic acid and polycytidylic acid, usually known as poly I:C. This had been found to be a very efficient inducer of interferon formation both in cell cultures and in laboratory animals. The findings from Hilleman's team were published as a series of articles in leading scientific journals.

Hilleman was not only a brilliant scientist but also an excellent businessman. In the following years he gave his observations wide publicity in lectures given at international meetings. He also wrote numerous review articles which discussed their implications in relation to the treatment of viral diseases and malignancies. His disappointing experiences with the production and purification of interferon were now clearly reflected in his attitudes. He said that he saw no future for exogenous interferon itself; instead, inducers were the magic which would soon open the way to the application of interferon for medical purposes.

The importance of these new findings was soon recognised around the world, and had a great influence on the course of interferon research for many years. I suspect that almost every laboratory in the world with interest in interferon started at least some work with poly I:C. This substance was very much in vogue, and it sucked scientists into the world of inducers. Fashionable follies can be dangerous in all fields, including science. I wonder how many years in the development of interferons were lost because of this craze for inducers, until it was generally realised that this particular path led to nowhere.

The initial goal of those working with inducers was to find some nucleic acid or polynucleotide which could be used as a drug in humans. Many problems arose. Poly I:C, though a highly effective inducer in cell cultures and animals, proved inactive in man because human blood contains an enzyme which breaks it down. But when at last, a more stable form of poly I:C was developed which induced interferon even in human body, it was found that this, like all related chemicals was quite toxic in man. The synthetic chemists found that they could not separate the toxicity from the ability to induce interferon: these always went hand in hand. With hindsight, one can see that in part, the adverse effects of the inducers were due to the large amounts of interferon which they evoked. However, not all their deleterious effects could be accounted for in this way.

Another problem was that, after the first use of an inducer, the body could not produce a second crop of interferon when the inducer was administered again after a short time. The same phenomenon which Kurt Paucker and I had observed in cell cultures, was found in experimental animals given an inducer. "Recovery" after the first induction took about a week, and only after this could an animal respond with full production of interferon when given the inducer again.

When the first inducer papers from Hilleman's laboratory appeared in 1967, a decade after the discovery of interferon, they completely eclipsed the results obtained with interferon by Finter, Gresser and ourselves. For some years thereafter, very few remained interested in the production and purification of interferon, for these topics seemed very remote from the main stream of activity. Even Tom Merigan's research interests switched from interferon to studies with interferon inducers.

My own attitude was not affected by Hilleman's work. It suited me well to work in peace, and I still had faith in the potential of white cell interferon.

Maurice Hilleman was a big man and a charismatic personality. He was intelligent, quick and aggressive. I recall a panel discussion in a congress organised by the New York Academy of Sciences in June 1969 in New York. The topic was the use of interferon and interferon inducers as medicines. The protagonist for interferon was Norman Finter; the advocate for inducers, the mighty and awe-inspiring Maurice Hilleman. Norman Finter began his talk by telling in Oxford English a story of how tigers are hunted in Bengal. A rope and a sheep are needed for the job. The hunter goes into the jungle, ties the sheep with the rope to a tree, and sits up in a tree with his gun, waiting for the tiger while the sheep bleats plaintively. He continued: "Sitting here next to the redoubtable Dr. Hilleman, I can easily imagine the feelings of that sheep! But remember, the object of the exercise was to shoot the tiger."

Hilleman's reply was quick as lightning: "My mother taught me as a child to beware of wolves that come dressed in sheep's clothing." I believe that, in the following discussion, he convinced many in the audience — but not all — of the superiority of inducers over interferon.

It must in fairness be added that the work of Hilleman's group on inducers represented research of a very high scientific standard. As he worked for a pharmaceutical company, it was natural that he should try to get the maximum publicity for his findings. It was not entirely his fault that so much attention in the following years was focused on inducers, and that work on the production and purification of interferon was largely disregarded and belittled. The course of the events rather reflected mass psychology and the relative lack of independent thinking amongst scientists.

During this period, I designed for some lecture a picture entitled "The better way to the clinical use of interferons." In the picture, a truck loaded with poly I:C is approaching the inducer express way which is already filled with heavy traffic. Nearby on the narrow and winding interferon road, there are only two pedestrians walking, while cows graze by the road side (these symbolised our finding that milk can be substituted for serum in the production of white cell interferon, which I shall mention later). The picture reflects my irritation and ironic feelings at the time. The popularity of interferon itself was at rock bottom, but I still had a pretty clear concept of "the better way." I was confident that the situation would eventually change and the side road would be widened to become the main highway.

22

The Yardstick

The activity of an interferon is usually measured in terms of its capacity to block the growth of some virus. In practice, this is done in the following way or by some minor variant. A series of dilutions are made of the interferon to be tested, and each is added to several identical cell cultures. These are incubated overnight to allow time for the interferon to produce its effects. Other cultures are left untreated with interferon, and serve as controls. On the following day, all the cultures are infected with a virus, chosen because it normally grows in the cells concerned and destroys them. If a culture has been treated with sufficient interferon, the cells are protected against the virus-induced destruction. The activity of the interferon is expressed in units, where one unit corresponds to the smallest amount of interferon, i.e., that present at the highest interferon dilution, which protects half of the cells against virus attack.

Although the principle is quite clear, the details of the methods used in particular laboratories are often very different, and their sensitivity varies greatly according to the exact method, and especially, the particular cell and virus used. This caused many problems during the first decade of interferon research. It was difficult to compare the results obtained in one laboratory with those in another, as there was no common "yardstick." In order to compare their results, many investigators therefore began to exchange interferon samples with others. In this way, it could be established, for example, how many "Merigan units" were present in one "Cantell unit." When such comparisons disclosed great differences in the units, it became obvious that some generally accepted and, preferably, internationally recognised standards needed to be established.

In England, the Scientific Committee on Interferon began to look into the matter, and one of its members, Norman Finter, requested interferon samples from a number of laboratories, including mine. In early 1967, the results of their comparative assays gave us some pleasing information: our own leukocyte interferon was about hundred times more potent than the interferon from Chany's laboratory, and compared very favourably with interferon made by Finter and Merigan. This information began to circulate among the inner circle of dedicated workers of interferon and was very helpful to us. The quality of our work and of our interferon began to assume a greater credibility in the eyes of these other scientists.

I had met Finter at a congress in Fort Lauderdale, Florida, late in 1966, and then offered to supply the British committee with interferon for the studies they planned in man. In their few earlier, and largely negative human volunteer studies, they had used monkey interferon, but the committee members now wanted to use human interferon in their future studies. After comparing interferons from different sources, the committee decided to accept my offer, and this launched a long-lasting collaboration between this interferon committee in England and my own laboratory.

The committee needed a standard for human interferon, i.e., an interferon preparation with a precisely determined and thereafter defined potency, which could be sent to laboratories around the world. By comparison with such a standard, amounts of interferon measured in different laboratories could be converted to a common unitage. The ultimate goal was that workers should always express amounts of interferon in terms of a common International Unit, based on such a standard: this would require the involvement of the World Health Organisation, which alone can authorise international standards.

The Scientific Committee on Interferon had already had samples of white cell interferon from Chany, which had been assayed in several laboratories. It was put forward as a provisional interferon standard, but its very low interferon content posed a problem: some laboratories using relatively insensitive assays could not register any activity in it. When it was found that our own leukocyte interferon was much more potent, the committee requested ten litres of our interferon so that a new standard could be prepared. I sent what was required to London, where it was distributed into several thousand small ampoules, and converted into a powder by freeze-drying: in this form, its activity was almost indefinitely stable when stored in the refrigerator. The material was given the code designation 69/19, and samples were sent to many laboratories around the world as a standard to be used when assaying human interferon.

The details of the procedures involved in preparing a reference standard for a biological material are incredibly complex, obviously important, but in essence, very dull. Furthermore, personal interests, ambitions and biases tend to become involved, as well as the many national and international organisations concerned with matters of standardisation. One of these organisations is the International Association of Microbiological Societies which has set up a permanent biological standardisation section. In the 1960s and 1970s, the driving force in this section was Frank Perkins, an Englishman with a very clear and determined mind. In late 1969, Perkins organised an international meeting in London to consider the standardisation of interferon and inducers. The meeting attracted considerable interest, and 84 people representing 13 countries participated. The goal was to agree on the choice of Standards or Reference Preparations for the various interferons prepared from different animal species, so that in future, all laboratories could work with each of these interferons in terms of the same unit. Many workers gave presentations during the three days of the meeting, and there was much lively discussion even though this often strayed into irrelavant topics. Perkins controlled the discussion with good humour but firmly, at times raising his voice to pull the meeting back to the main issues.

The outcome of the meeting was that some significant advances were made in standardising interferons. Chany's interferon preparation was adopted as a Research Standard, and assigned a potency of 100 human interferon units per ampoule. It was known at that time that our interferon preparation, 69/19, was considerably more potent, but results were not yet available from all the various laboratories that had taken part in an international collaborative assay study. Soon after the meeting, all the results came through, and on the basis of these, an ampoule of 69/19 was defined as containing 5000 interferon units. To Charles Chany's disappointment, our 69/19 preparation replaced his material, and has since been used generally as the standard for human interferon around the world: it was later officially recognised by the WHO as the International Standard for human leukocyte interferon, and has been the starting point for the standardisation of many other different human interferon preparations since.

23

Competitors

We were not the only ones to study the production of leukocyte interferon during the 1960s. Similar work was carried out by Charles Chany in Paris, Spencer Lee in Canada, Fred Wheelock in the U.S. and their colleagues, as well as a few others, but in all these laboratories, only small-scale research studies were done. Apart from us, only academician Soloviev and his team in Moscow aimed at the large-scale production of interferon for clinical use. On several occasions, I received interferon samples from Moscow; each time I found our own product to be about one hundred times more potent.

Charles Chany's group soon gave up work with white blood cells, and they, and also Rita's group in Italy, looked instead at the production of interferon from cells of human placental membranes. It was an interesting idea, as placentas, being a "by product" of childbirth, are readily available. Unfortunately, neither group persevered for very long with their studies, and after they had published their results, they carried the work no further. Thanks to the excellent Finnish Red Cross Blood Transfusion Service and my established post as a government employee, I was in an exceptionally fortunate position. Nevertheless, it seems strange that development of large scale human white cell interferon production was not addressed seriously in any other laboratories. Perhaps the task was considered too uncertain, or too laborious and too dull. But even though there was little competition in my chosen field, there was no shortage of critics.

The lack of competition suited me particularly well. I have always steered clear of popular fields, and I would never have been successful in frenzied competition with laboratories with big teams and highly sophisticated techniques. My preference has been to paddle along gently in some quiet, uncrowded backwater and at my own chosen pace.

The major competition with our white cell interferon seemed more likely from another quarter, namely from fibroblast interferon. During the 1960s, it had been shown that fibroblasts, i.e., cells from human connective tissue, grow readily in the laboratory, and during culture they retain the characteristics and properties of normal cells. This made them very much more attractive as a possible source of interferon than cells of human continuous lines: these latter could be grown readily in the laboratory, and often be subcultured repeatedly over many months or apparently even indefinitely, but they originated from cancer tissue, or had become transformed into cancer-like cells during culture.

It is true that, in spite of their normal characteristics, fibroblasts have a theoretical limitation: they are only able to divide about 50 times in culture before they show signs of senescence. Once this stage has been reached, the cells grow much more slowly, and then soon die (this in itself an exciting phenomenon which is now being studied in relation to the problems of aging). Theoretically, however, huge numbers of cells are obtainable after 50 cell divisions: by then, each single cell would produce 10^{15} descendants, i.e., a thousand million million. Such calculations show that the finite life span of fibroblasts would not in practice be a bottleneck as far as mass production of interferon is concerned.

Human fibroblasts are readily obtained from two sources, from abortion tissues, and from the human foreskins obtained after circumcisions, a routine procedure for baby boys belonging to the Jewish faith. Both fetal and foreskin fibroblasts grow excellently in culture.

Tom Merigan in Stanford was the first to try to make interferon from foreskin fibroblasts, and he published a paper on this subject in 1966, the same year as that of our first publication describing white cell interferon production. In their paper, Tom and his co-workers suggested that their procedure could provide interferon in amounts large enough to test as an antiviral agent in man, but they did not undertake the laborious task. The reason I have already mentioned: like so many other interferon scientists, Tom became swept up in the general enthusiasm for interferon inducers, and he gave up further studies on production. Nevertheless, some other groups started to work on the production of human fibroblast interferon, notably the group directed by Pieter De Somer at the Rega institute in Leuven, Belgium, and that of Jan Vilcek in New York. Both worked out an elegant method for increasing the yield of fibroblast interferon, and on a small scale, obtained good yields. Unfortunately their attempts to scale up the work soon ran into difficulties.

To grow fibroblasts in the laboratory is relatively laborious and expensive, because these cells will not grow in suspension in the nutrient medium in a culture flask; instead they must attach to the glass or plastic surface of the culture container before they can divide and multiply (they are for this reason called anchorage-dependent cells). To grow many cells therefore requires a correspondingly large surface, and because of this, it proved technically very difficult to produce fibroblasts in very large numbers. In consequence, the amounts of fibroblast interferon made were much less than the amounts of leukocyte interferon we produced.

Those working on fibroblast interferon laid great stress on the theoretical merits of their product, especially its safety, which they contrasted with that of leukocyte interferon. They pointed out that the amounts of the latter will always be limited by the availability of fresh human blood. They highlighted its potential dangers, the possibility that viruses present in the donor blood might contaminate the interferon and be transmitted to and cause disease in recipients. They also suggested that white cell protein contaminants in the leukocyte interferon might cause allergic reactions.

Hans Strander and I were much irritated by such comments. We were of course fully aware of all these possibilities, for we had discussed them time and time again, and we were sure that these potential problems could be overcome. We intended to proceed with great caution, and if indeed we ever reached the stage of contemplating the use of our interferon in patients, we would ensure that all these risks were reduced to the absolute minimum, if indeed they could not be eliminated completely.

Something which particularly annoyed me was that these criticisms directed at leukocyte interferon seemed to have a strong influence on the attitude of those in blood transfusion centres outside Finland. The result was that the use of white blood cells as a rich source of human interferon was largely ignored for over 10 years. While this may not have been what our critics intended, in practice this was exactly the outcome. As I saw it, the demand for interferon looked potentially so enormous and the difficulties of making it so formidable that all possible avenues should have been explored. Furthermore, it gradually became apparent that the interferons obtained from white cells and from fibroblasts were by no means identical. Rather than being competitors, they were two different types of human interferon, and both might find uses.

24

Against The Current

In the talk I gave in Siena in 1967, I discussed the possible outcome of our leukocyte interferon production method, which was then beginning to take shape after our four years of development work. I concluded with the following sentence: "If crude leukocyte interferon could be purified without great loss of activity, the clinical usefulness of interferon could be put to a critical test." I wanted to leave the audience and those who later read my article with a positive picture of the capabilities of our method, but in my own mind, the word "if" in this last sentence highlighted the crux of the problem. I was certain we could make hundreds of litres of crude interferon, but they would be of no value if we could not purify and concentrate them. This made me feel very insecure and indeed rather pessimistic.

At that time, interferon had been under study for 10 years in many parts of the world, but very little progress had been made in relation to its purification or in analysing its chemical composition. The leader in this field was a chemist from Czechoslovakia, Karl Fantes, who had gone to England as a refugee in 1939, and worked for the pharmaceutical company Glaxo. He was a member of the Scientific Committee on Interferon, and had worked intensively on the purification of chick interferon.

His best preparations contained about one million units of biological activity per milligram of protein. Nevertheless when he analysed these by a technique, electrophoresis, which separates molecules in terms of their electric charge, he found two proteins, which formed separate bands on the electrophoresis gel. Both these proteins proved to be biologically inert and therefore merely impurities: all the interferon activity lay in the region between them where there was no detectable protein. Karl showed in this way that even his best preparations were still far from pure. Thus pure interferon must have a biological activity even greater than had been expected, and indeed higher than that of most other known biological substances. Furthermore, his studies led to some uncertainty about the nature of interferon. Was it really a protein, or could it after all be a carbohydrate? Karl also showed that chick interferon was not a single substance but a group of similar substances, and the procedure used for their purification looked very cumbersome.

The obvious conclusion from these and other purification studies was that with only the resources in my own laboratory, there was no chance of succeeding in the purification of leukocyte interferon. I had trained as a physician and my knowledge and experience in biochemistry were modest. The only hope was to collaborate with some expert in this specialised field. I had links with interferon scientists in England, which dated from talks with Norman Finter and Karl Fantes on the beach at Fort Lauderdale in Florida during the conference there in 1966. They were members of the Scientific Committee, which aimed at carrying out studies in man with human interferon. I offered to supply leukocyte interferon for their purpose, but of course this would need to be at least partially purified. Karl set to work to apply the tricks he had found useful when purifying chick interferon, but it soon became clear that these did not work with our interferon, and new ones must be devised. Within two years, he worked out a simple and effective precipitation method for concentrating leukocyte interferon with good recovery, but it led to very little purification.

At the 1969 Interferon Symposium in London, Karl described his interferon purification studies, and concluded that the day when pure interferon would be available was still far off. Even more disappointingly from my point of view, he said that chick interferon seemed a more suitable candidate than human interferon for the purpose of working out the basic structure of an interferon. So it seemed that the British committee was only interested in purifying our interferon to an extent sufficient to make it acceptable for studies in man. For my own more ambitious purification target, I would need to find another collaborator. Joseph Sonnabend seemed to be the man for the job.

Joseph came originally from South Africa, and had worked in the team of Alick Isaacs in London before moving to the United States. He knew a competent protein chemist by the name Panayotis Katsoyannis, and got him interested in interferon purification. He managed to obtain funding for the work from the U.S. National Institutes of Health (NIH), but unfortunately, the project came to a standstill very soon. Purification turned out to be so difficult that my partners lost interest. However, the collaboration marked the beginning of long-continued financial support for our studies by the NIH, and more specifically, by one of its constituents, the National Institute of Allergy and Infectious Diseases (NIAID). This funding was of great importance both to me and to the Blood Center. I was delighted that, at long last, the Blood Center got some recompense for all its expenditure on my behalf. Also, the American money enabled us to expand the production of interferon.

In 1969, the prospects for purifying our leukocyte interferon looked gloomy indeed. I remember vividly an incident at the end of October that year, during the interferon symposium in London. There was a conference dinner, and I sat next to Tom Merigan. During the meal, I told him about our problems with interferon purification, and Tom listened but made no reply. I turned to talk with Hilton Levy, an enthusiast for inducers from the NIH, who sat on my other side. After a while, Tom nudged me and said: "Kari, perhaps the answer is in your hands." I was taken aback and asked him what he meant. He replied: "It may be that you will have to work out the purification method yourself." Surprised and a little irritated, I told him we were hardly likely to be successful when highly trained and competent biochemists had so far failed. Nevertheless, his comments were very important in the sense that they forced me to re-evaluate the situation, and sowed the seed of the idea that we must ourselves tackle purification, however difficult the problems seemed.

The fact that our crude interferon preparations contained human serum made their purification much more difficult, but if it was not present in the medium, the white cells yielded relatively very little interferon. Hans and I had therefore tried hard to find substitutes for serum but with little success. Together with Gyurka Hadhazy and Hans, I tried to find out why serum was needed, and here we did succeed in getting some clues.

In 1969, Charles Chany organised an international interferon meeting in Lyon. I gave a paper on the need for serum in the production of leukocyte interferon, but my talk seemed to have little interest for the audience. This was hardly a surprise, for inducers were then very much in vogue, and to most people, our work seemed remote from the main stream of interferon research. In the middle of the meeting, Aila phoned from Finland with terrible news: my sister Maija had committed suicide. I was overwhelmed with grief and with anxiety for the fate of Maija's small children, and I left immediately for Finland, telling only Kurt Paucker of the reason for my abrupt departure.

We continued trying to unravel the serum problem with Dorothy Tovell, a new post-doctoral fellow from Canada who spent two years in my laboratory. I kept going over and over in my mind what might substitute for serum in the production of interferon. One day an idea came to me and it happened in a rather banal way.

Every afternoon at about 5 o'clock, all the family had dinner together in the kitchen of our home in Maunula. This was an important occasion in our life, during which the members of the family exchanged all the latest news. I watched my three sons eating hungrily, and washing their food down with gulps of milk (something which I never do, for I have a lactose intolerance, and never intentionally put milk or milk products into my mouth). Suddenly the thought struck me like a clap of thunder: my goodness, milk is a cheap biological fluid which we have never tested as a serum substitute. The next day, I went to the supermarket in Ruskeasuo, near the laboratory, and purchased various milk products, and with my co-workers, Sinikka Hirvonen and Dorothy Tovell, I started to experiment with them. The results proved absolutely conclusive: white cells cultured in the presence of milk produced as much interferon as they did in the presence of serum. In further studies, we soon showed that the active component in milk was its main protein ingredient, casein.

When later, we made leukocyte interferon for trials in patients, we did not dare use casein rather than human serum, for fear that it might evoke allergic reactions in the recipients. However, our results with milk were published, and Sidney Pestka's group in the U.S. made use of them when later they worked on the purification of leukocyte interferon.

The hope that we might be able to carry out trials with our interferon in patients gave a strong impetus to our purification work. My co-worker Hans Strander returned to Stockholm, and there he completed his medical studies, and presented his work in Finland on interferon production in his doctoral thesis. Hans decided to follow the career of a clinician and to specialise in oncology. We kept in close touch and he visited Helsinki frequently, staying with us. We all enjoyed Aila's delicious meals and our sauna, and over drinks, we talked about interferon until the early hours.

Hans spent his days with cancer patients, and so we naturally planned to test our interferon in patients with malignant conditions: studies with cell cultures and animals had shown very clearly that in addition to its virus-inhibitory actions, interferon had numerous effects on tumour cells and could inhibit the growth of tumours in animals. No doubt, if Hans had specialised in infectious diseases, these would have been the first clinical targets, but as it was, we tried our interferon first in malignant disease.

Karl Fantes had shown us how to concentrate our interferon, and as a result, we had considerable quantities of preparations containing about one million interferon units per millilitre, by far more potent than any other human interferon then available. We knew this was nevertheless still very impure, though not the extent of its purity. From present knowledge, it seems likely that our preparations at that time contained only about 0.01% interferon protein: the remaining 99.99% consisted of "impurities," most of these being the proteins in the human serum added during their production. As these proteins are naturally present in everyone's circulating blood, we were not greatly concerned that they were present in our preparations intended for clinical studies.

We made careful plans for trials with our interferon in cancer patients in Stockholm, and submitted them for approval to the Ethical Committee of the Karolinska hospital, where Hans worked. In his application to this Committee, Hans carefully drew attention to all the possible risks involved. The Committee gave approval for the study, and in 1969, Hans gave the very first injection of our interferon to a patient in the cancer clinic (the Radiumhemmet) at the Karolinska Hospital.

This first patient had a malignant skin tumour, a melanoma, which originated from a birthmark and had spread all over the body. Hans injected 450 000 units of our interferon into one of his veins. We chose this route of injection in the hope it would give the interferon its greatest chance to come into contact with the cancer cells all over the body. About two hours after the injection, the patient had an unexpected and violent reaction. He became nauseated and feverish and began to vomit, to have chills and to shiver intensely, and his blood pressure and pulse rate increased. Hans gave him an injection of cortisone which alleviated the symptoms, and after 10 hours, these had all disappeared. We were intensely disappointed. Was this going to put an end to our ambitious plans to use interferon in cancer treatment? And why had it happened? We had given a very high dose of the same interferon preparation to rabbits, and it had caused only a slight rise in body temperature and no other symptoms.

25

The Lesson From Animal Experiments

At that time, we were also carrying out many studies with our interferon in animals. While the aim was to follow the fate of the interferon after an injection, we were at the same time naturally very eager to see whether it caused any adverse effects. The answer to this second question was very clear cut. We injected bigger and bigger doses, but even gigantic doses of our leukocyte interferon caused no detectable side effects in any animal species studied. In my institute, we gave interferon to mice, rats, guinea pigs, rabbits and sheep; the injections and subsequent collection of blood samples were skilfully performed by our animal house biologist, Liisa Pyhälä.

In collaboration with workers in other countries, our interferon was injected into monkeys belonging to a variety of different species. In none of these animals did our interferon cause any discernible adverse effects even when injected directly into the blood stream. Why then had it caused the severe reaction in Hans' patient? Was it due to the species specificity of interferons, the fact that one works best in the cells of the particular species from which it is derived? But human interferon was known to be active in monkey cells and vice versa, and yet even when injected intravenously, our interferon had not produced symptoms in monkeys, but only in the patient. There seemed to be something strange going on.

At that time, virtually nothing was known about the fate of injected interferon. Earlier experiments in mice in other laboratories had shown that when injected directly into the blood stream, interferon disappeared very quickly from the circulation, and after some ten minutes, none remained detectable there. Because our interferon was so potent, we could inject doses that were hundreds of times higher than those in these previous studies. We found in our experiments that the injected interferon was again rapidly cleared from the blood at first, but then the rate at which it disappeared slowed down greatly, and some interferon could still be detected in the blood even 24 hours after the injection.

What would happen if we were to inject our interferon into the muscles or into the layer beneath the skin, the subcutaneous tissue? We experimented in rabbits, giving high doses of our interferon by intramuscular or by subcutaneous injection. I felt quite sure that we would not detect any interferon in the blood, but the results were startling: within 15 minutes from the time of the injection, we could detect interferon in the blood without difficulty, and it remained there at a stable level for the next 24 hours.

We did similar experiments with animals from many species. These showed that an injection of 200 000 interferon units per kilogram of body weight led consistently to a concentration of about 100 units per millilitre in the blood of them all. These results fired my imagination. In the animal studies up till then, I had somehow regarded interferon as something unreal, mysterious, labile and intangible. The new results showed that when a sufficient amount of interferon was injected into animals, it behaved in the same regular and consistent way one would expect for any other drug.

Our findings had other direct practical implications. There was no need to inject interferon directly into the blood of patients, and indeed it seemed preferable to give it subcutaneously or intramuscularly. This would be much simpler and safer, and seemed likely to result in better and more long-lasting effects.

26

Eureka!

Research usually advances in small steps which often result from some chance event, and progress is made via many detours. Our purification efforts were significantly affected by a casual remark which I described in the chapter "A Possibility" in the beginning of this book.

David Habif was Professor of Surgery at Columbia University in New York and his speciality was breast cancer. During his long career, he had seen how often surgery and chemotherapy had proved unsatisfactory treatments for this common form of malignancy. This was why this skilled surgeon became so greatly interested in interferon during the late 1960s. Habif had numerous very wealthy patients and friends, and while he had no great interest in money for himself, he made very good use of his connections to raise money for research. This enabled him to establish a laboratory in New York where Samuel Graff, a retired professor of biochemistry at Columbia University, could try to produce interferon in HeLa cells. I had used the same cells in my first studies of interferon and shown, contrary to Isaacs' theory, that they can indeed be induced to make interferon.

Sam Graff soon learned how to grow HeLa cells in suspension culture, and the cells grew freely in the nutrient medium. However, his attempts to make interferon ran into difficulties, which David

Habif described to me when he made his first visit to my laboratory in 1971. He also explained his own hopes for using interferon therapy in patients with breast cancer, and I in turn described what we were doing. This visit marked the beginning of our collaboration and warm friendship, which lasted until David died 20 years later.

Sam Graff brought up the possible use of ethanol (the ingredient in all alcoholic drinks) for the purification of interferon during his first visit to my laboratory in February 1972. I never asked later why this idea came to him. Was it an intuitive insight, an educated guess based on his long experience as a biochemist, or just pure guesswork? Likewise, I cannot explain why, out of all the other possibilities I discussed with Sam, I became interested only in trying ethanol. However, for some reason, I found the idea of using ethanol particularly attractive. My goal was to purify interferon especially for use in patients. Therefore, it was of utmost importance to use safe reagents, so that if any residual traces were left in the product, they would not be harmful. From this point of view, ethanol, cheap, safe and "almost physiological," was an ideal reagent. Might an additional reason have been that I do not mind fortifying myself with an occasional glass of spirits, even though I am a son of a teetotaler.

I started to try the use of ethanol for the purification of our interferon with my co-worker, Sinikka Hirvonen, who was a pharmacist. She had joined the department of virolgy while I was in Philadelphia, and took charge of the mump vaccine production, which she did with great competence. I began to involve her more and more in interferon studies, and she acquired a taste for research in general, and into interferon in particular. Over the years, she developed into an excellent co-worker, and her organisational skills proved invaluable when we expanded the production of interferon. I learned to trust her absolute reliability, and to make use of her practical experience and preciseness when planning experiments, and of her skilful hands in their execution.

As a first step, we precipitated interferon by a method we had adopted from one described by Karl Fantes. We then dissolved the interferon in 1/30th of its original volume in a buffered salt solution, and tested the effect of ethanol on its activity. High concentrations of ethanol were not harmful to interferon provided the temperature was low. Interferon dissolved in ethanol under acidic conditions, but was precipitated when the pH rose toward neutrality. Sinikka made very accurate notes, and a scheme for the purification of interferon began to evolve. From results we had obtained, the next experiment was designed and the work proceeded in small steps.

The process began by precipitating the interferon in a crude preparation. We then discovered that interferon could be extracted directly from the precipitate by mixing this in an apparatus like a kitchen blendor in cold 94% acidic ethanol. The next step was to raise the pH very slowly, all the time stirring the solution and keeping it in an ice bath. As the acidity decreased, proteins began to precipitate. We found that the first to precipitate were impurity proteins, which could be removed by centrifugation. The interferon itself only precipitated when the pH was raised even nearer to neutrality, and it could then be redissolved, precipitated again, and concentrated, just as at the beginning of the process.

After much work, we gradually came up with a method which enabled large volumes of interferon to be purified about hundredfold. We had called our earlier concentrated preparations C-IF (i.e., concentrated interferon), and so we termed the new preparations P-IF (for purified interferon). Our P-IF preparations were still impure — later work showed that they contained only about 1 per cent interferon — but they were in quite a different class from our C-IF preparations. So P-IF replaced C-IF in our laboratory, animal and eventually clinical studies.

No doubt in the eyes of a chemist, our purification procedure looked quite unusual. Eva Vahvaselkä, one of the protein chemists in our institute, shook her head with disbelief when I told her about our activities with ethanol. But I was not a chemist, and my ignorance gave me freedom from prejudice. The whole procedure was worked out little by little as a result of numerous trials and errors.

A new post-doctoral fellow, Erik Mogensen, came from London to join my laboratory. Erik was born in Denmark and was still a Danish citizen although he had lived most of his life in England and written his Ph.D. thesis in the University of London. Perhaps it was his Danish origin that motivated him to come to work in Finland. He spent two years in my laboratory and then moved on to join that of my friend, Ion Gresser, in Paris.

It was a pleasure to work with Erik. He was a skilful chemist with original ideas, and his diligence and integrity impressed me. His doctoral thesis had dealt with the culture of mouse cells, and the production and purification of mouse interferon. The original plan was that he would study interferon production in my laboratory. However, I asked him first to look into the stability of our two interferon preparations, C-IF and the brand new P-IF. This had considerable practical importance for us and it also had theoretical interest.

At that time, human interferon had the reputation of being very unstable. This stemmed from the studies of Jan Vilcek, Bill Carter and their co-workers in the United States, and Karl Fantes had made similar observations with our interferon. According to these investigators, human interferon became increasingly labile as it was purified more and more. I was deeply concerned by this prospect especially now that we had made significant progress with purification, but my own limited experience was quite the opposite to what had been suggested by these others.

In collaboration, Erik and I soon showed that our P-IF preparations were a lot more stable than the less pure C-IF preparations. Indeed, P-IF was astonishingly stable: it could even be held for 3 days at 56°C without loss of activity. Also two reagents, guanidine hydrochloride and urea, which are well-known to unfold the three-dimensional structure of proteins, did not irreversibly inactivate our interferon; on the contrary, they actually afforded it protection, and in their presence, our interferon could even be boiled for a short time. We also found that a detergent, called SDS, which binds tightly to proteins, did not reduce the activity of our interferon. We showed that our white cell interferon contains at least one disulphide bridge — this means that there are two regions in the interferon molecule that are bound tightly to each other by a chemical bond between sulphur atoms.

Altogether, the work with Erik showed that far from being a labile substance, human leukocyte interferon was a very robust molecule which could tolerate rough treatment. Why then had the other laboratories previously found such different results? Even today, I do not understand why our interferon appeared to lose activity in the experiments carried out by Karl Fantes. On the other hand, there is a logical explanation for the results reported by the other laboratories, for in these, fibroblast interferon had been studied. It is now known that human leukocytes and human fibroblasts produce different kinds of interferons. Fibroblast interferon, which has since been rechristened beta interferon, is much less stable than leukocyte interferon, which is now called alpha interferon. How stupid and short-sighted it had been to set leukocyte interferon and fibroblast interferon up as competitors and to make claims and counter-claims about the superiority of each.

Sam Graff, now looking a very old man, revisited my laboratory with his wife and co-worker Ada in August 1972. I told them about the progress of our purification studies, and these charming people and my family were together to celebrate my 40th birthday. My own happiness was subdued by the obvious melancholy of my guests, which was for reasons which only became apparent later. Sam's work had not progressed well, and his sponsors had decided to close down his laboratory. In addition, his persistent cough proved to be due to lung cancer, and he succumbed to this disease after a short time. I still remember with much gratitude the suggestion which Sam made in February 1972 even though its full significance did not become apparent to me until much later.

During its purification, interferon is held for several hours exposed to a high concentration of acid alcohol. This is the stage at which any viruses which might be derived from the blood of a donor are very effectively destroyed. Due to the harsh treatment used during their purification, our interferon preparations have never transmitted any virus infection to a recipient as far as I know, and this is course of the utmost importance in the present era of AIDS. This safety factor introduced by our use of ethanol helped much later to enable leukocyte interferon to be registered as a drug in Finland and elsewhere.

In May 1973, the Tissue Culture Association of the U.S. organised a meeting on "The Production and Use of Interferon for the Treatment and Prevention of Human Virus Infections." This was held in Lake Placid in New York. I was invited to review our work on leukocyte interferon which had been then in progress for ten years. I looked at the magnificent mountain scenery through the window of the auditorium and thought of my countryman, Veli Saarinen, who had here won the Olympic gold medal in the 50 km cross country skiing race in 1932, the year of my birth.

In my lecture, I described, for the first time, the principle of our purification method. The paper was published in the following year in the series called In Vitro. We have since made numerous small modifications to the procedure, but its basis has already been established at that time and has remained unchanged.

The story goes that Archimedes cried out in triumph "Eureka", meaning "I have the answer", when in a flash of inspiration, he perceived for the first time the nature of the physical phenomenon which has since been called the law of Archimedes. The word Eureka has since been used to describe a similar sudden new insight into an apparently intractable problem. Every scientist has the joy of experiencing similar but usually small "Eurekas" during his career, and I have already told of some Eureka experiences from my own career, such as the sudden idea of using milk in interferon production.

There is a book by Carl Grimberg that is famous in Scandinavian countries called "The History of The Nations." It contains a vivid description of the excavations sponsored by Lord Carnarvon, and carried out in the Valley of The Kings in Egypt by Howard Carter during a period of eight years. The results were meagre, and in 1922, it was decided not to continue the work after the end of the coming winter. However, at the beginning of November, the discovery was made of a flight of stairs cut in the rock, and this was the start of a dramatic series of events.

On November 26th, excitement reached its climax when a sealed door was reached. With shaking hands, Carter made a small hole in its upper left corner, and as he pushed a candle in through the opening, its flame flickered as air gushed out. It was some time before his eyes adapted to the dim light, and then, struck dumb with amazement, he watched as shapes began to emerge from the darkness. Lord Carnarvon could stand the suspense no longer and asked anxiously: "Can you see anything?" — "Wonderful things" was all that Carter could falteringly reply. They had found the tomb of the pharaoh Tutankhamon, who had died while still very young. This immeasurably important discovery has given a vivid picture of life in Egypt three thousand years ago. This story, more dramatic than fiction, has long fascinated me, and in my imagination I have often shared with Carter these exciting moments.

In the early years of the 1970s, I began to form a general concept from the results of all our efforts over the previous ten years. It consisted of several parts. We had learned how to handle huge numbers of white cells and how to obtain consistent production of interferon from them. The Blood Transfusion Service had joined in the exercise, and the prospects for large scale interferon production looked excellent. In spite of all my pessimism stemming from the anticipated difficulties, we had begun to crack the problem of purifying our interferon, and could now purify it a hundredfold and concentrate it a thousandfold even on a large scale. We had obtained quite unexpected information which showed that our interferon was quite remarkably stable. This made it easy to handle in the laboratory; furthermore, it could easily be sent anywhere in the world in a simple postal package. We had found that when interferon was injected intramuscularly or subcutaneously into a variety of animals, it soon entered the circulating blood; with daily injections, a more or less constant blood concentration of interferon could be attained.

As I assembled all these points in my head, I began to realise that although our work had been criticised, belittled and ignored by so many, our efforts had indeed been worthwhile. I realised that I now had information unknown to others, and also interferon of a quality not available to anyone else in the world. I could see what possibilities lay in the future. I knew I was the only human being who knew all this and grasped its significance. This indeed was a great "Eureka" experience for me. I realised, nevertheless, that our main goal was to apply interferon for the treatment of sick people. Would all our progress in the laboratory prove beneficial in the clinic?

27

From The Laboratory To The Clinic

After ten years of research, interferon still remained merely something with promise for the treatment of virus infections and cancer. In the opinion of many people, however, that promise was certainly not great, and perhaps not even small; attitudes towards its use as a medicine were pessimistic in the extreme, or totally negative. There had been early enthusiasm after the discovery of interferon, but this had led nowhere. The embers had been fanned into life with the discovery of the interferon inducers, but these too had been a disappointment — indeed in the following years, they slowly disappeared from the scene, their silent departure contrasting strikingly with their noisy entrance. Overall, interest in the medical application of interferons and inducers reached its lowest point in the early 1970s.

The thing which continued to hamper clinical studies with interferon was the simple but chronic lack of preparations for testing. Without these, one could only sit at one's desk and theorise about the merits and disadvantages of interferon. These could not be put to the test in medical practice.

In increasing numbers, representatives from various pharmaceutical companies paid visits to my laboratory. The industry was clearly interested in interferon, but no company was prepared to launch a long-range project to produce it. The outcome apparently seemed too uncertain and the investment too risky. There were in fact at least two possibilities open to them, to make interferon from leukocytes or from fibroblasts, but neither was utilised. My impression was that many companies were closely following developments in the field, and some even had small interferon programmes running in their laboratories so as to be able to move more quickly if there was a breakthrough. But how could anything really important come forward if everyone waited passively for someone else to move? Somebody had to break the vicious circle. Somebody had to make interferon suitable for studies in patients. This was the goal I set for my laboratory.

By the end of the 1960s, we began to have sufficient quantities of potent but very impure interferon to start small-scale studies in patients. But would this interferon be safe enough to inject into patients? We knew we would have to give it by injection, for like others, we had shown in animal experiments that after oral administration, interferon was all destroyed in the digestive tract and none entered the blood stream. So we did not even try giving interferon to patients by mouth.

There were a number of risks potentially involved with giving interferon by injection. In practice, the greatest of these was the possibility of transmitting a hepatitis virus. There are several different viruses that cause hepatitis (infections of the liver), and it was known that blood or blood products could sometimes transmit these diseases.

Little was then known about hepatitis viruses, except that they were known to be extremely stable and resistant to inactivation by chemical means. So it was not completely impossible that our concentrated leukocyte interferon might contain an infectious virus. We knew that our interferon tolerated highly acid conditions quite well, and in order to destroy possible blood-borne viruses, including hepatitis viruses, we routinely treated our interferon at pH 2, i.e., under very acidic conditions. While this did not give a total guarantee of freedom from infectious viruses, we gained additional comfort from the fact that, in contrast to many other countries, blood-borne hepatitis infections are very uncommon in Finland.

I aimed to be involved with studies in patients in two fields. My collaboration with Hans and his Swedish colleagues had as its goal the treatment of cancer. The collaboration with the scientists in the English Committee would test interferon in the prevention and treatment of certain virus infections. Since the members of this Committee were particularly concerned about the risk of hepatitis, The Blood Transfusion Service put into place elaborate extra arrangements to select and screen the blood donors so as to minimise further the risk of hepatitis.

Another potential danger was that our interferon might cause severe allergic reactions in patients, or even in the worst case a life-threatening anaphylactic shock.

On the other hand, it had to be remembered that in the Swedish study, the plan was to administer interferon to patients who were severely ill from some form of cancer. The hope was that they would gain benefit from the treatment, and the potential benefits and risks had to be weighed against each other as in any medical treatment.

Hans Strander described the planned studies in an application which he submitted to The Swedish Medical Board and The Ethical Committee of the Karolinska Hospital. Both gave their permission.

I have mentioned earlier what happened when Hans gave an intravenous injection of interferon to the first patient in 1969. We did not understand why the patient developed such severe reactions after the injection. Fortunately soon afterwards, results from the animal experiments gave the welcome indication that the injections should be given intramuscularly rather than intravenously. I told Hans about our results and suggested that for the next patient, he should inject the interferon into the muscles. In the next few days, I waited with excitement for his news: would there be such a violent reaction after intramuscular interferon as after an intravenous dose? Hans first gave a very small dose and then gradually increasing the doses during the following days. To our great delight and relief, nothing dramatic happened, even when Hans finally raised the dose to 3 million units. Even the original melanoma patient who had had such a severe reaction after the intravenous interferon, tolerated intramuscular interferon well.

Something happened though. All patients given interferon developed fever which lasted 6 to 12 hours, and they experienced influenza-like symptoms. This was surprising and interesting, but the important thing was that these side effects were so mild that they did not prevent interferon therapy. Indeed, as compared to the side effects caused by most other cancer treatments, these adverse effects of interferon were minor and well tolerated.

The significance of these results was very clear to us. We had finally reached the point where it was possible to test interferon in patients. A wide new ocean was open for us to explore.

Hans took blood samples from patients who had received interferon and sent them to Helsinki. We measured their interferon content and found that the intramuscularly injected interferon had behaved in patients in exactly the same way as in our animal experiments. In man too, interferon was transported slowly but efficiently from the injection site to the circulation; after even a single injection, the blood level of interferon remained fairly stable for many hours.

I went to Stockholm to see the patients who had received interferon. I still vividly remember one, a young man who had a melanoma. His situation was hopeless, for the malignant tumour, which had originated from a birthmark on the skin, had spread to many parts of his body. The calm and friendly behaviour of this pale, slim, dying boy, and the ray of hope which he glimpsed in interferon, has stayed with me since, a haunting memory.

I had received training as a physician, and working in a cancer hospital, I had become very interested in malignant disease. Interferon had first attracted me from the point of view of its basic science, but its use in medicine had gradually risen to become the main aim of my efforts. The first series of patients consisted of 11 patients with various forms of cancer. Three of the patients also suffered from shingles, a painful and occasionally dangerous infection caused by herpes zoster virus. They received amounts of interferon ranging between one and five million units per dose, amounts chosen on the basis of experiments carried out with our interferon in England. These had examined how much interferon was needed to protect monkeys against experimental virus infections, and their results suggested that a dose in the order of a few million units of interferon might protect humans at least against some virus infections. In our patients, such injections caused some malaise and fever, lasting for about 12 hours. The patients were given three injections weekly, and during the course of treatment, the side effects subsided. In some patients, the treatment was continued for more than 6 months.

There were no dramatic changes in the tumour in any of the patients. Those who had shingles recovered from the virus infection, but it was by no means certain that this was due to interferon. So the main message from this first study was that patients could be treated with moderate doses of interferon for long periods of time. We were also encouraged by the fact that there was not a single case of hepatitis or any allergic reactions among our patients.

We sent our first paper describing these studies in patients to the American publication, the *Journal of The National Cancer Institute* in December 1972. When the paper appeared the following year, we noticed something annoying. The paper had as its title "Systemic Administration of Potent Interferon to Man." The journal printed papers in several different sections, one of which was called "Clinical and Laboratory Investigations on Man", and we wanted our paper to appear in this section. This is what in fact happened, but because of some misunderstanding, the paper appeared with the compound title: "Clinical and Laboratory Investigations on Man: Systemic Administration of Potent Interferon to Man." This senseless and awkward title still grates whenever I read it, and this has been often, for this paper has since been referred to innumerable times in the scientific literature.

Now that Hans and I knew we could examine the efficacy of our interferon in patients, we spent much time trying to decide on which diseases we should concentrate our investigations. In spite of all our efforts, we still had only very limited amounts of interferon available, and it would be inadvisable to spread our efforts too widely. On the other hand, we did not want to put all our eggs into one basket.

Hans visited Helsinki frequently, and we kept in close contact by phone calls and correspondence. Finally, after much discussion, we decided to test our interferon in four malignant diseases, acute leukemia, Hodgkin's disease, myeloma and osteosarcoma. Thinking about our decisions later, I realise that our reasoning was too much influenced and restricted by an overemphasis on the oldest known activity of interferon, its ability to inhibit virus multiplication. We then still reasoned that any effect of our interferon in cancer would result from its antiviral activities. We therefore chose these particular four malignant diseases, because there were some suspicions indicating that viruses of some type were their cause, even though there was no definite proof. Perhaps I should have known better, for I had been in the team which had discovered the cytostatic property of interferon, i.e., its ability to inhibit cell division. I had also read many accounts of the numerous and varied biological activities of interferons. Nevertheless, being a virologist, I laid too much emphasis at this time on the antiviral actions of interferons.

Hans carried out some limited studies in leukemia and Hodgkin's disease. Six children with advanced leukemia were given interferon injections for a couple of months. No definite conclusions about their effect on the leukemia could be drawn from the results, but the attending physicians drew attention to the fact that none of the children contracted a virus infection during the time they were under treatment with interferon.

The condition of a 24-year old man with Hodgkin's disease improved at the beginning of his course of interferon treatment, but after 6 months, the disease began to progress again even though interferon therapy was continued. Some other patients with Hodgkin's disease, and others with myeloma showed similar promising but unsustained benefits during interferon treatment. This marked the beginning of a long series of important studies on the use of interferon for the treatment of myeloma which still continue in Sweden and elsewhere.

Overall, however, osteosarcoma became the most important subject for the studies which Hans carried out with interferon. This malignant bone tumor, which occurs in children and adolescents, has a very poor prognosis. It is, fortunately, a relatively rare disease, but one that runs a most dramatic course. The disease begins usually in one of the long bones of the arm or leg. At that time, in spite of radical treatment, including amputation of the limb and local radiotherapy, it usually terminated fatally within a year or two. The real problem was the metastases, the secondary tumours which all too often resulted from cells which spread through the blood from the primary tumour and lodged in the lungs and other organs. Removal of the primary tumour, however soon this was done after the first diagnosis, was usually too late to prevent the spread, and it was the secondary tumours which finally killed the young patients. Treatment with the cytostatic drugs which were then available was of doubtful value.

Hans gave daily injections of 3 million units of interferon to his osteosarcoma patients for a month, and during this time, the tumour was removed by surgery. The patients then went home but continued to receive interferon injections three times weekly for 18 months. During this period, the injections were given sometimes by a nurse, sometimes by the parents, and sometimes by the patients themselves. The results obtained in the first few patients looked so promising that at the end of 1971, it was decided to treat all the osteosarcoma patients admitted to the Karolinska Hospital with interferon. This was a bad mistake as I realised, but only several years later — it is so easy to be wise after the event. What we should have done was to treat every second patient with interferon and to leave every second patient as a control in a "double-blind" study of the type which I shall describe later, in the chapter with that name. We did not carry out such a study for what then seemed very good reasons: the early results looked promising and raised our hopes of seeing a major clinical benefit from interferon therapy; the total number of patients coming to the hospital was quite small, so that if only alternate patients had been treated, results would have accumulated even more slowly; and we hoped that the results would be convincing even without proper controls.

In the event, the results were good but not good enough, and indeed the value of interferon therapy for osteosarcoma still remains an open question. But notwithstanding this, these osteosarcoma studies which Hans carried out had a great impact on the future of interferon research, as I will relate later.

Sometimes an observation on a single patient can have far-reaching importance. This was the case in one patient that Hans studied in the early 1970s. She had had large and painful warts on the soles of her feet since childhood. While she received interferon therapy, the warts disappeared, but they recurred some 6 weeks after the interferon treatment was completed. Hans grasped the significance of this chance observation for he knew that warts are caused by viruses, even though little information about the causal agents, the papilloma viruses was at that time available. To treat ordinary harmless warts of the skin with interferon did not look sensible, but there is a wart disease, called laryngeal papillomatosis, which is most troublesome, and sometimes even life-threatening. In this disease, which usually occurs in children, cauliflower-like tumours grow in the region of the vocal cords. At first they merely cause hoarseness and problems with speech, but in the later stages of the disease, they may seriously interfere with breathing or even lead to suffocation unless a tracheotomy is carried out.

The laryngeal papillomas can be treated surgically, but, unfortunately, they usually recur rapidly. In consequence, the small patients may have to undergo scores of operations. The tumours are benign, but on rare occasions they may become malignant. Under the microscope, the papillomas resemble warts in structure, and virus particles are sometimes seen. They are caused by a papillomavirus which is probably transmitted from the mother's genital tract to the child during childbirth.

When it was observed that plantar warts healed when a cancer patient was receiving interferon therapy, a trial was started in Stockholm in which children with laryngeal papillomatosis were given interferon injections. The results were encouraging in that, during the interferon treatment, the tumours in most patients diminished in size or disappeared completely. Unfortunately, they slowly reappeared after the cessation of treatment, and it proved difficult to achieve a permanent cure. Therefore, treatment was continued in many patients for years, interferon being given in the smallest dose and at the minimum frequency needed to keep the papillomas under control. In general, the children tolerated the interferon injections well, and had fewer side effects than adult patients.

I have already described our work on the purification of leukocyte interferon. In the beginning of the 1970s, we began to have our partially purified P-IF preparations to try alongside the cruder concentrated interferon C-IF preparations. When Hans began to work with P-IF, he liked it immediately. The problem with C-IF, beside its low purity, was its low activity. We could not concentrate the crude interferon more than 30-fold, at most, and the resultant preparations contained about one million units of interferon per ml, often a bit less. We wanted to give large interferon doses to our patients, but the volume of C-IF then needed became a problem. For example, to give 3 million units to a patient daily, at least three millilitres had to be injected into his or her muscles. Such large injections caused some local pain which was greater according to the total volume.

At the start, we packaged our interferon in the glass Wassermann tubes which were in general use in the laboratory, and sent it airfreight to Stockholm. Hans kept on asking that the volume of the interferon to be injected should be reduced in some way; his most commonly used dose, 3 million units, was largely determined by the volume of the impure interferon, for the patients would have tolerated higher doses, but for the local pain caused by an even larger volume. So he greatly welcomed the P-IF preparations for with these, the volume was no longer a problem. We dispensed P-IF in 0.5 millilitre aliquots in small plastic tubes. Three million units still remained the daily routine dose, but with the introduction of the P-IF preparations, local reactions became a thing of the past.

We had cherished the hope that perhaps with P-IF, even the systemic side effects would diminish or disappear, but here we met with disappointment. Injections of P-IF still caused fever, fatigue, chills, headache, etc., the symptoms resembling those of influenza. A collaborator of Hans, Snorri Ingimarsson from Iceland, took as the subject for his doctoral thesis the analysis of the side effects of interferon. We began to suspect more and more that the adverse effects were not due to impurities, but to the interferon itself. Later, when completely pure interferons became available, this suspicion was confirmed.

The clinical studies with interferon showed that certain of the symptoms of influenza and other virus diseases are actually caused by the interferon which is produced in the body during the infection. From the point of view of the organism, these may well be useful. For example, fatigue and fever will persuade the patient to rest, and this will hasten recovery from the infection.

With the use of P-IF, we confirmed our earlier observation that patients rapidly become partially tolerant to the adverse effects of interferon. The greatest symptoms are seen with the very first injection, but they become much less pronounced after a few further injections. Nevertheless, the fact remains that interferon therapy is associated with some unpleasant side effects, and it is necessary for the patient to learn to live with them. Later on, I got direct experience of these effects when I received interferon injections in connection with a research project.

28

Double-Blind

To prove that a new drug is efficacious is not an easy task. Many diseases, for example acute virus infections, usually get better without any treatment being given. As for most chronic diseases, it is very difficult to predict the course in an individual patient. For example, even in patients with the same type of cancer, there are great differences in rate at which the disease progresses. Lung cancer may lead to the death of one patient only a few weeks after it is first diagnosed, while another patient with an apparently similar tumour may live for several years.

If a novel drug has a dramatic effect on a severe disease, its value is easily shown. Penicillin, for example, was such a drug, especially when it was first introduced: it saved patients from otherwise lethal bacterial infections, so that there was no possible doubt about its activity when only a few patients had been treated. Such drugs are rare exceptions among the hundreds that are newly introduced into medicine each year.

Much more commonly, a new drug has, at best, only a slight advantage over others which are already in use. As the benefits obtained in the treatment of individual patients are likely to vary considerably, results drawn on the basis of trials in only a few patients may give little or no indication of its true efficacy. Often, the apparently promising preliminary results obtained in a small pilot study are not confirmed in subsequent more extensive trials. Indeed, it is usually necessary to enter large numbers of patients into carefully designed and controlled clinical trials, and to follow their progress for long periods before firm conclusions about the value of a new drug can be drawn.

This situation seems one not easy for a layman to grasp. All too often, a patient, a relative, or a friend becomes convinced of the efficacy of a new remedy on the basis of a single case — his own or someone known to him. Not infrequently, this situation is exploited by the unscrupulous as a way of making money. Nevertheless, between the two extremes of clear proof and outright fraud, there is a twilight zone of uncertain information in which many kinds of charlatan and, sadly, even some physicians operate. These believers include not only "simple Simons" but also highly intelligent and otherwise knowledgeable individuals. Linus Pauling, the winner of two Nobel prizes, represented someone in the latter category. On the basis of his personal experiences, he believed firmly in the efficacy of vitamin C in the prevention of colds in spite of all the scientific evidence to the contrary.

Critical evaluation of a new drug proceeds in the clinic in stages. First, the acute adverse effects and the dosage have to be defined, and only then is it time to study its efficacy. The results obtained can be compared with those obtained previously with other treatments, but such historical controls may be unreliable for a number of reasons. Therefore, patients are usually divided in two groups, one being treated by the best current medication and the other by the new drug. Such a comparison will yield more reliable results.

However, even this experimental design has its weaknesses. Wishful thinking and biases among the investigators and other staff may influence the way in which patients are allocated into the two treatment arms and the results are evaluated. To eliminate these human factors, the following design should be used. A number of patients, who are as similar as possible in terms of age, sex, type and stage of disease, etc., are chosen for the study, and they are randomly allocated into two groups according to some pre-arranged system. Both groups receive medication in coded but otherwise identical vials. Half are given the test drug, and half a placebo, that is to say an ineffective mock preparation. In this way, the investigators, the hospital staff, and the patients do not know which vials contain drug and which placebo. The code is not broken until the study is finished, and only then can the results be analysed. A study of this ideal type is referred to as a placebo-controlled, double-blind trial, and is the best way, indeed usually the only way, to obtain reliable information about the true benefit of a new treatment.

Such a study is not always possible: for example, it is not ethical to give a placebo to one group, if there is already an effective treatment for that disease. In such a case, the new treatment must be compared with the old one. However, it is still often possible to carry out the study in a double-blind fashion, so that neither the investigators nor the patients know which treatment is given to a particular patient.

When tested in one of these controlled ways, it is often found that a suggested new treatment for some form of cancer is completely ineffective. Nevertheless, such a negative result is well worthwhile. It settles an issue, and enables the investigator to move on to a new problem. The worst study is one which because of defects inherent in its design cannot give any definite and reliable new information, and to carry out such a study is not ethical.

29

Interest In Interferons Increases

Professor Pieter De Somer in Belgium was one of the pioneers of interferon research. He was the head of a research laboratory, the Rega Institute, in the University of Leuven, and he also founded and directed a successful pharmaceutical company. Later he became the president of the University. He became interested in interferon soon after its discovery, and collected a team of active young scientists to study it. One of the main goals of his team was to make interferon from cultures of human fibroblasts and use this for studies in patients. Alfons Billiau became the leading figure in this project.

In September 1971, De Somer and his colleagues organised an international interferon congress in Leuven. At this time, Hans had been conducting clinical studies in Stockholm for some time, but it was too early to speak about the results. Nevertheless, some information about the ongoing studies had already leaked out, and some of the participants in the meeting asked about his progress.

I talked about the use of milk and its main protein, casein, in the production of leukocyte interferon. The results were very pleasing to me, but attracted little interest among the audience. The Belgian workers extolled the merits of their production system, but I could see that we were far ahead of them; we had much more interferon available and our clinical studies were in full swing.

In Leuven, I met for the first time an interesting man from Texas, William E. Stewart II, or briefly Bill, and I subsequently followed his eventful career for nearly 20 years. At this first meeting, he wore hippie dress and his long hair was tied behind with a bow, in pirate fashion. He was very relaxed and self-assured in his behaviour, and was accompanied by a young wife and a small child. He had come from the U.S. to work in De Somer's laboratory, and this unconventional Texan scientist seemed to be getting on well in Belgium. He made a number of very pertinent comments during discussion periods at the meeting. I observed him with some slight surprise and amusement.

Tom Merigan, as quick-witted as ever, had soon come to realise that at least the current generation of inducers were after all not very promising. He changed camp again, and refocused his interest on interferons. We had been in correspondence before the meeting, and in Leuven we discussed possible clinical studies to be carried out with our interferon at his university, Stanford, in California.

Tom was particularly interested in testing interferon for the treatment of shingles in cancer patients. At Stanford, numerous patients with lymphomas, i.e., malignant tumours of the lymphoid tissue, came for treatment. Many developed shingles, a disease caused by the virus that causes chicken pox in children. When a child recovers from chicken pox, the virus remains dormant in the body for the rest of life, but may become active again and start to grow if the natural resistance of the body is lowered for some reason. The virus then does not produce chicken pox, but instead shingles, or girdle pain, or as it is technically termed, herpes zoster. Zoster is a serious condition in cancer patients, because it may spread widely to involve not only different areas in the skin but also the internal organs, and it may cause the patient's death. Recovery from zoster is often followed by local pain at the site of the former shingles site. This pain may be very severe and persist for many months after the skin blisters heal.

Tom had had much experience in the medical treatment of zoster, and his plan for testing interferon in this condition sounded excellent. He also planned to use interferon to treat children with cancer who developed chicken pox, for this is life-threatening in these patients. Preliminary studies in Sweden had not yet shown whether interferon was effective in herpes zoster, and it seemed likely that Tom's well-controlled studies would yield reliable information. Tom was well respected at National Institutes of Health in Washington, and he was quite confident of sufficient financial support from that source to cover the cost of any interferon we supplied to him. The studies in Sweden which had shown the safety of our interferon in man were a great help in getting approval for the Stanford trials, and these were soon underway.

In the early 1970s, I began increasingly to be in touch with scientists in Japan. I corresponded with Professor Tsunataro Kishida of the University of Kyoto, and Dr. Koichi Kato of the large industrial organisation, Toray. I sent them both samples of our interferon and information about our latest results, and they both visited my laboratory. Kishida and I wrote a short review of current interferon research which appeared in 1974 in Japanese in a publication series called Saishin Igaku.

Kishida spent long periods working in Paris. He visited Finland with his family and we became friends. Although his English was not of the best, we got along splendidly. He had a bright mind and a wonderful sense of humour. In Finland, we went with our families to the Ateneum art museum, and saw there a modern work in which the onlooker's image was multiplied many many times by mirrors. Kishida's remark "many Professor Cantell" continued to cause amusement to my little sons for a long time afterwards.

As is the Japanese custom, Kishida was an enthusiastic and skilful photographer of people, a "homo photographicus." I heard that while working in Pasteur Institute in Paris, he went to much trouble to assemble all the Institute staff for a group photograph. He set the camera on an automatic exposure timer, and then ran back to join the group. Unfortunately, however, he had aimed the camera too low, and all those in the picture appeared only as headless bodies, apart from the tiny Kishida himself.

In the spring of 1974, I received a letter in French from Professor Yasuiti Nagano, the doyen of interferon research in Japan, inviting me to attend an interferon symposium to be held in the following year in Kyoto, in connection with the General Assembly of the Japan Medical Society. The speakers would be Nagano, Nagata, Kishida and Cantell. I accepted the invitation with pleasure, and numerous letters were exchanged between Helsinki, Tokyo and Kyoto to settle the details. In the event, my planned itinerary was disrupted to some extent by the unexpected intervention of a lady called Mathilde Krim.

Krim was educated in Geneva, and pursued her doctoral studies there in genetics before moving to the Weizmann Institute in Israel in the 1950s. She married a U.S. lawyer, Arthur Krim, who was a member of the board of that Institute, and went with him to the United States of America. Arthur was the president of United Artists, a film company founded by Charles Chaplin, and a very wealthy man. Mathilde now had a house in Manhattan, a summer home on Long Island and a ranch in Texas. Arthur was an influential member of the Democratic party, and Presidents Kennedy, Johnson and Carter had all visited the Krim home.

At the beginning of the 1970s, Krim was a member of a national committee which produced a report concerning the state of the art in cancer research, and prospects for the future. As a result, she became familiar with the literature of interferon and aware of its great potential as a treatment for cancer. She saw to her dismay what meagre funds were being devoted to interferon research in the U.S. and decided she would try to change this. As a first step, she began to plan an international meeting to which some 60 interferon scientists would be invited, together with industrial representatives, and officials concerned with the allocation of government and private funds for the support of scientific research.

In spite of letters and phone calls from Mathilde, I was at first reluctant to take part in her meeting, because it partly overlapped with my long-planned trip to Japan. There was, however, an other reason for my hesitation: I felt uncomfortable and inclined to be cautious when faced with Mathilde's drive and evident political manoevering. Although Hans had obtained results that were encouraging, we were still at the very beginning of our studies, and as yet had no definitive proof that our interferon was effective in a clinical condition. Would the meeting do us more good or harm? Nevertheless, faced with Mathilde's persistence, I decided in the end to go to New York, and fly on from there to Tokyo immediately after the meeting.

I set off on Sunday, March 30 for London where Hans and Ulf Nilsonne, the surgeon who had operated the Swedish osteosarcoma patients, joined me, and we boarded a plane for New York. From previous journeys I had learned to dress comfortably for a long flight, and so was wearing a casual sweater when we arrived in the evening only to find that my luggage had gone astray, a not infrequent event when travelling by air. Thoroughly annoyed, I had to go on to the hotel with just my briefcase. Krim had invited all the participants to a party at her house that evening, and so I had to go there still wearing my sweater. Mathilde and Arthur Krim greeted me cordially and I admired their large and beautiful home, which contained many interesting objects including personally signed photographs of several U.S. Presidents, Oscar statuettes, reflecting Arthur's connection with the film industry, and many other items.

When I conversed with some of the others present who were participants in the meeting, I was surprised by their attitudes. There seemed to be considerable suspicion, criticism and, overall, a negative view of Mathilde, or even open hostility. These impressions were strengthened during the days of the meeting. This was attended by representatives from the National Institutes of Health (NIH) and its subdivision, the National Institute of Allergy and Infectious Diseases (NIAID), institutions which distributed large sums of money for research; the latter had provided funds to the Finnish Blood Center in connection with my collaboration with Tom Merigan. These, and staff representing various U.S. governmental agencies were particularly antagonistic in their stance in relation to Krim.

I do not know why so many people found Krim so irritating. Perhaps it was due to the strong influence that she had on the way in which federal money was allocated to research, and that this was perceived as a threat by the biomedical establishment, or that she seemed to be meddling in area which was felt to be no concern of hers. Perhaps her character, wealth, political connections and gender played a role, or it seemed she had more enthusiasm than critical judgement or expertise. Perhaps above all, her growing influence was feared.

From my point of view, much more disquieting was my growing impression of a surprisingly malicious, suspicious and negative attitude directed towards our work in Finland and Sweden in particular, and towards interferon in general. Before the meeting, Hans had sent to some U.S. pathologists at their request detailed information about his osteosarcoma patients, and samples taken from their tumours. I heard rumours that these doctors would claim that many of the tumours were not osteosarcomas, and had been wrongly diagnosed by the pathologists in Sweden. I became worried and angry. I knew of course that there are several kinds of osteosarcomas, each running a different clinical course, and that distinction between them is not simple, but I had complete confidence in the skill of the Swedish pathologists. Nevertheless, I feared that if my worst fears were fulfilled, there might be attempts to humiliate Hans and me in public and in effect accuse us of fraudulent claims; at the same, this would crush all further clinical use of interferon. It was of course true that the results Hans had obtained were still very preliminary, and based on a comparatively small number of patients. However, it had been an honest open investigation. We had concealed nothing and had yet not attempted to draw any firm conclusions. Would nevertheless the hostility and envy directed at Mathilde Krim be focused on our work, and so on interferon research in general? Such thoughts nagged at me during the night, and I slept miserably.

The next day, I sought out Tom Merigan, for I knew he had good connections with Washington. I wondered whether perhaps he also had some reservations about this rapidly convened meeting; he was himself organising a big interferon congress at the end of the summer in Stanford, which might prove an anticlimax if all the key new findings in interferon research had already been made public in New York. I told Tom of my concerns and added that I would discontinue all collaboration with any scientist in the U.S.A., if our study was trashed in public on unjustified grounds. In view of my anxiety, Tom immediately telephoned Alan Rabson, a leading pathologist in Washington, who was coming to the meeting but had not yet arrived. He gave Tom reassuring information: for the most part, there were no discrepancies in the diagnosis given to our patients. I felt somewhat relieved.

In the end, about two hundred people took part in the meeting, which was held in the auditorium of Rockefeller University. I described what we had done in relation to the production and purification of our interferon, and our animal experiments with it. I said that we had made 50 000 million units of interferon during the previous year in Finland, using for the purpose the white cells from 70 000 blood donors. I stressed that this had been a tremendous undertaking and it would be rather negative to consider the total result as merely some tens of milligrams of pure interferon (from present knowledge, the correct figure was 250 mg); on the contrary, this emphasised the quite extraordinary biological activity of interferon. I compared and discussed the various ways in which interferon might be produced in bulk for clinical studies. Later, I chaired one of the sessions and acted as the coordinator of the discussions in another.

As the meeting proceeded, I got the impression that the ice began to melt: the general atmosphere became more relaxed and attitudes less aggressive. Shortly after the meeting, Tom Merigan wrote to say that, in his opinion, the meeting had been an enormous success and had given a great boost to research on interferon.

When the Krim meeting finished, I left immediately for Tokyo, and from there travelled with my friend Kishida in a bullet train to Kyoto. The symposium there seemed to go well, and Nagano and Kishida were both satisfied. The Japanese Ministry of Health began to make plans for the production of leukocyte interferon in Japan, and I met members of the Japanese Interferon Committee. I also visited the Green Cross company in Osaka which was involved in interferon studies, and there gave seminars and participated in discussions. I had carried out some joint work with Professor Kuwata of Chiba University, and he gave me as a present a very fine copy of a Hokusai woodcut. This fine collaborator has since sent me some additional pictures by Hokusai and Hiroshige, which now decorate our living room and have brought much pleasure to all my family. All my hosts were overwhelmingly friendly and hospitable during my stay in Japan, and I found Japanese food, especially sushi, very much to my liking. Altogether, I greatly enjoyed my first visit to Japan.

I flew back via Moscow, and at last I could begin to relax after so much tension. The weather was really beautiful. I drank champagne, watched the endless steppes, mountains and rivers of Mother Russia through the window, and listened several times to Beethoven's violin concerto. With my strenuous trip behind me, I now felt immense relief and satisfaction, and I recall these happy hours each time I hear this wonderful piece of music.

30

Things Start To Move

The meeting which Mathilde Krim organised in 1975 had far-reaching effects on many aspects of interferon development. Interferon began to come to the attention of many influential people. While many retained a critical or hypercritical attitude towards it and its clinical potential, there was an increasing number who were now interested in it or even cautiously favourable. Mathilde produced a report of the meeting which she used as a weapon to get her way, and she set her sights in particular on those who could control the allocation of federal funds to support interferon research. Up till then, the National Cancer Institute (NCI) had provided no such support, and its directors, Frank Rauscher and Vincent DeVita, did not understand interferon's promise. Nevertheless, because of Mathilde's pressure, the NCI allocated one million dollars for interferon research in 1975. In terms of their total budget, the sum was very modest, but it was an important beginning. Some of this money was used to purchase interferon from the Finnish Blood Center, and the rest was divided among several projects in the U.S., including support for Tom Merigan's lymphoma trial.

The main emphasis of Mathilde Krim's meeting had been on the potential uses of interferon in cancer, but its use in virus infections was also covered, and the meeting helped to accelerate developments also in this area. The National Institute of Allergy and Infectious Diseases (NIAID) had already provided funds for Merigan's study on shingles and chicken pox. Now they increased their financial support, and he began to study interferon as a treatment for patients with hepatitis. This marked the beginning of an extremely important line of research.

Chronic infection of the liver with the virus of hepatitis B, was, and still is, a common and important disease, which often leads to cirrhosis of the liver and liver cancer. In 1976, Merigan's group published the results of a trial of Finnish leukocyte interferon in the treatment of four patients with chronic hepatitis B. The treatment was unquestionably beneficial in three, but with the particular treatment regime used, it did not permanently cure the chronic infection.

One of the leading figures in the NIAID was Dr. George Galasso, who had himself worked with interferon, and, thanks to his good offices, interferon was used in many good clinical studies. Even before Mathilde's meeting, I had a letter from Marty Hirsch in Boston, who was studying patients who received a kidney or bone marrow transplant. It was necessary to treat these patients with immunosuppressive drugs so that their body did not reject the transplant, but because of these drugs, many developed severe virus infections. Hirsch had the idea, in my opinion, an excellent one, that interferon treatment might prevent such infections. I was delighted when the NIAID gave support for the study; the interferon needed was ordered from the only possible source — the Finnish Blood Center.

Two participants in the Krim meeting had a strong impact on the development of interferon research in later years. One was Jordan Gutterman, who worked in the largest cancer hospital in the U.S.A, the M.D. Anderson Hospital in Houston, Texas. Gutterman had previously studied cancer immunotherapy, especially the effects on tumours of repeated vaccination with

BCG (the Calmette-Guerin vaccine). This particular avenue had led nowhere, but he remained greatly attracted to the idea of treating cancer by stimulation of the body's own defence mechanisms. He saw the hidden potential of interferon, and began to explore how he could get his hands on supplies. During the meeting, Gutterman asked me whether it would be possible to obtain some of our interferon. I have no recollection of our talks — I met so many people during those few days — but I told him, like many others, that there was, unfortunately, only enough interferon for a few to use, and that, off-hand, I could give no promise. However, he was so determined to get interferon that he was not discouraged.

Another participant at Mathilde's meeting was a remarkable elderly lady, Mary Lasker. Her husband, Albert Lasker had headed an advertising bureau, and one of their ideas had been the slogan "LSMFT = Lucky Strike means fine tobacco," which entered the conscious and subconscious minds of millions of American smokers, and in return, brought millions of dollars to Lasker. Although neither had any training in medicine, he and Mary established a Foundation to support medical research, particularly into cancer. Thus ironically, a substantial part of his wealth, acquired by promoting one of the greatest causes of cancer, now became directed to its prevention and treatment.

Mary Lasker had money, influence, insight and the reputation of being an philanthropist. She was much respected and she had contributed greatly to the development of the National Cancer Institute of the United States. She became very impressed with the potential for interferon — apparently, her attitude was influenced by the fact that a close friend had breast cancer that was not responsive to drug treatment. Interferon became an important link between she and Jordan Gutterman, when in August 1977, Mary Lasker decided that her Foundation would purchase one million dollars worth of Finnish interferon for use in his clinical studies. At the same time, Gutterman learned that he was to receive some of the interferon purchased from Finland with federal money. His efforts during the previous two years began to yield a harvest.

In October 1977, I gave a lecture with the title "Human leukocyte interferon for clinical use" at the meeting held to celebrate the 500 years anniversary of the University of Uppsala in Sweden. Jordan Gutterman was anxious to talk to Hans and me, and because we were all so busy, we agreed to meet at Arlanda airport, near Stockholm, on October 12th. Gutterman was accompanied by Mrs. Deeda Blair, who represented the Lasker Foundation. We talked about the design of Gutterman's research plans, and I gave a preliminary promise that he would get the interferon he needed. I flew back to Helsinki, while Gutterman spent the following day with Hans in his hospital in Stockholm. Gutterman was particularly impressed by a patient with myeloma who had unquestionably benefited from interferon therapy. He decided he would try interferon in patients with advanced forms of cancer.

Early in 1978, Gutterman finally started to use interferon in his patients. His studies went along fast, and only a few months later, he was able to report at an interferon meeting organised by the NIAID in Bethesda that breast cancers in some patients had shrunk in size during interferon therapy. At the same meeting, Tom Merigan reported similar findings in patients with lymphoma.

Gutterman's enthusiasm for interferon and his faith in it were so greatly strengthened by his early results that, with the active encouragement of Mary Lasker, he took a bold step: he applied to the American Cancer Society for a grant of two million dollars to purchase more of our interferon for his expanding clinical trials. The newly elected president of this Society was Frank Rauscher, previously the Director of the National Cancer Institute. His attitude toward interferon had been totally negative and he had consistently blocked all attempts to divert NCI funds towards interferon research. Whether the latest results or his new position softened his attitude is not clear, but Rauscher invited a panel of ten experts to New York to evaluate Gutterman's application. These supported the proposals, and the American Cancer Society gave the biggest grant in its history, the two million dollars requested. The news of this large purchase of interferon was quickly spread through the mass media and triggered an avalanche of interferon-related publicity. As will be seen, this was both helpful and harmful to the cause of interferon research.

31

Is It Difficult To Grind A Barrel Organ?

More and more visitors came to my laboratory as stories about interferon appeared increasingly in the press. By the end of the 1970s, they were so numerous as to become quite a nuisance for me and my small team.

I have always adopted an open doors policy. I did not try to patent any of our findings, and I published details of all our methods and of improvements as soon as they were established. I had no wish to make personal profit out of interferon, and my wife fully supported me in this decision. I have never regretted this policy, and it does not worry me that many others have made fortunes as the result of our efforts. My attitude was also influenced by the fact that I was a state employee, working in a national institution and financed largely by public funds. Furthermore, my work was entirely dependent on blood given voluntarily by Finnish donors. It may sound sentimental and perhaps even priggish to say this, but I genuinely felt in return that it was my duty to work hard and make all my knowledge freely available to other investigators, because it might help with the treatment of sick people.

I often discussed these matters with Harri Nevanlinna and our ways of thinking were identical. When clinical studies became a reality and the demand for leukocyte interferon increased sharply, the Blood Center began in 1971 making it. In the following year, the new building of the Finnish Red Cross Blood Transfusion Service opened, and this enabled their output of interferon to be considerably increased. In October 1972, a graduate chemist, Hanna-Leena Kauppinen, was engaged to supervise the preparation of interferon in the Blood Service. This was the beginning of a happy collaboration which now has continued for more than 20 years. I also had very close connections with Harri's "right hand" man in the Blood Service, Gunnar Myllylä, and I think with great pleasure of my work with him, and with Juhani Leikola, when later he took over from Harri.

To transfer our methodology to the Red Cross Blood Center took more time than I had expected. The procedures used in the production of leukocyte interferon look simple enough, but numerous small details affect the end result. It is easy enough to induce formation of some interferon by treating white cell cultures with Sendai virus, but the essential thing is to obtain the maximum yields and to recover as much interferon as possible at the end of the purification process. Most other laboratories that tried our procedures obtained for quite some time interferon yields lower than those in my laboratory. However, after about two years of work, the yields in the Blood Center reached our level. From then, and to my great relief, the Blood Center began to receive a part of the steady stream of visitors.

Our visitors came from all round the world — Nordic countries, all parts of Europe including the Eastern bloc, the United States of America, South America, South Africa, Japan, Australia. Most, especially the Japanese and the Americans, were astonished by the small and modest accommodation in our "factory." They were a very heterogeneous lot: there were serious research workers, fortune hunters and all sorts in between. From our point of view, the easiest visitor was the one who spent only an hour or two in my office and laboratories to hear the latest news about our own efforts and about interferon in general. The greatest disruption resulted from the occasional large delegation who wanted to go into all our procedures in minute detail. I began to try to limit the size of the groups, to take visitors from different countries at the same time, and to restrict the duration of their stay to one week at the most. I recommended those who studied our work seriously to make a second visit when they had gained some practical experience in their own laboratories.

All these visitors took much of our time, and, inevitably, our own work was affected to some extent. My staff consisted only of 6 or 7 technicians, and they became accustomed to carrying on with their work while the visitors watched and asked frequent questions. Usually, they spoke in English, and one result was that my co-workers became considerably more proficient in that tongue. Among the visitors, we soon identified two particularly irritating categories, the arrogant and the mistrustful, and with all the pressures on us, these sometimes stretched my patience to the limit. The arrogant visitor belittled our methodology and hastened to tell us how to improve it. It was difficult to convince such an individual that, for example, we had found more interferon was produced in round bottles and under certain precise conditions than in slightly different bottles and under slightly different conditions. We had optimised our production through literally hundreds of trials and errors. However, sometimes, such a visitor would come back to my laboratory in a much humbler frame of mind after having personally experienced many failures.

The title of this chapter relates to a story which I heard from my late father. An old organ grinder was asked scornfully by a listener whether it was difficult to play his "instrument." The man replied that it was not really very difficult but that nevertheless, he had learned certain tricks. "Such as what," insisted the hearer. "Well, for example," the old man said, "If you try to turn the handle in the wrong direction, you will find that it comes off in your hand." This is what happened to many of our visitors; they tried as it were to turn the handle in the wrong direction. Nevertheless, with persistence and sometimes after repeated visits, their efforts were usually finally rewarded with success.

Mistrustful visitors were fortunately few in number, for they were particularly irritating. They usually came from some pharmaceutical companies, and suspected that in spite of our apparent openness, we were keeping back some vital details, essential for the process to work satisfactorily. Such a person tried to set traps, to ask in turn the same question from different members of my staff and to thumb through our records and stocks of reagents and interferons.

We could not help but be influenced in our attitude by a visitor's personality. It was much nicer to offer helpful advice to someone who was pleasant than to one who was disagreeable. I found that my own opinions in this respect usually agreed well with those of my female co-workers. I will not mention the names of any of the offenders here, but among the many guests who were pleasant, I will mention in particular an Italian professor, Diego Balducci. He visited me many times and we became friends. With an admirable persistence, he overcame all obstacles, and transferred our methodology successfully to his company in Rome. Finally, he was able to register his product, "Alfaferone" in Italy, where it has become a successful drug.

From their attitude, I could see right away that many of our visitors would never succeed in copying our procedure. Thus unfortunately, for the most part, the time we spent in teaching them was a waste.

On several occasions, I had visitors from two laboratories in the former Yugoslavia, one in Belgrade and the other in Zagreb. I tried to persuade them that they should collaborate and so benefit from their mutual experience. Only much later, with the outbreak of civil war and the disintegration of Yugoslavia, did I finally realise how naive I had been in my ignorance of the terrible racial hatred between the ethnic groups in that region. Collaboration between them was obviously quite out of the question.

32

The Cubans

Many visitors have faded from my mind, but those from Cuba are associated with so many vivid memories that I shall relate some in this chapter.

At the beginning of 1981, the Cuban ambassador to Finland and a Cuban professor visited my laboratory and asked numerous questions about interferon. Clearly, the subject was of considerable interest to them, and, after carefully sounding me out, they asked if I would accept a team of visitors from Cuba to learn how to make and purify leukocyte interferon. This topic had just become the subject of a barrage of publicity in the media, and we were very busy in the laboratory. I replied that I would, but asked that the numbers in the "delegation" should be kept to a minimum, and the duration of the visit to a week. At the time, I was fairly sure that the visit would be a complete waste of my time, but I did not want to abandon our open doors policy.

On Monday March 30th, a group of six Cubans, virologists, immunologists and biochemists, headed by Manuel Limonta, a specialist in internal medicine, came to my laboratory. They were all tired and jet-lagged after their long journey, but they set to work without delay to try to understand our process; they followed our procedures in detail and took copious notes.

The ambassador invited Aila and me to a dinner which was followed by rum drinks and dancing to Cuban rhythms. However, when the ambassador conveyed to me personal greetings from Fidel Castro, and a request from him for interferon, I began to be suspicious. What was behind all this? Was Castro himself or somebody close to him suffering from cancer, and was this the reason for his keen interest in interferon? To be on the safe side, I decided to lock all the laboratory freezers which contained our interferon stocks. Afterwards, I realised that this was an exaggerated over-reaction. As the week proceeded and I started to get better acquainted with my guests, they began to rise in my estimation.

The team returned to Cuba, and at the beginning of May, I had a letter from Limonta to tell me that their interferon laboratory was nearly finished. On behalf of his government, he invited Aila and me to visit Cuba in July to see the laboratory and spend a vacation there. I was not keen to go, but in response to much coaxing by the Cuban embassy in Helsinki, I suggested that my co-worker, Sinikka Hirvonen, should go to Cuba. Sinikka was willing to go, but asked that her son Tapio could accompany her — he had also worked with interferon in our laboratory. The Cubans were happy with this suggestion, and she went.

When Sinikka returned, her story astonished me. Interferon production was in a full swing in Cuba in a laboratory converted from a former luxury house in a suburb of Havana. Furthermore, clinical studies with interferon had already begun in a virus infection called dengue. This is transmitted by mosquitos and there was a big outbreak in Cuba at the time. The local population was firmly convinced that this resulted from biological warfare launched by the United States of America, but I gave this story no credence. Nevertheless, the possibility of trying interferon therapy in dengue interested me greatly. Quite unexpectedly, I soon got some personal practical experience in this connection.

Sinikka had been bitten by mosquitos in Havana, and, soon after her return, she became severely ill with a disease which according to the textbook was a classical case of dengue. She suddenly developed a high fever, headache, and pain in her eyes and at the back of her neck, and two days later she came out in a blotchy rash. I gave her interferon injections and she soon recovered — whether the interferon played any part in this, or whether she would have recovered equally quickly without it, is impossible to say.

After repeated invitations, Aila and I decided to go to Cuba at the beginning of the following year. We were treated as guests of the government. We travelled in first class, lived in a beautiful white villa and were taken everywhere in a splendid black Mercedes-Benz limousine. We visited the "old" interferon laboratory which is now called Sinikka's house, and went to the inauguration of the brand new and elegant interferon institute which was adjacent. Just before the ceremony, I was informed that I was expected to cut the blue ribbon across the main entrance and to say a few words, which I duly did. Fidel Castro himself and several of the leading politicians in the country were present, and together with Castro, we toured the building, all the time followed by a TV film crew. Afterwards, we moved to a nearby villa and in the guarded inner courtyard, chatted with Castro while rum drinks were served. Security was very tight: no-one told us in advance that Castro would be there.

I gave two lectures on interferon in the Academy of Sciences to an audience which included several leading personalities. I was interviewed on TV. The hospitality was overwhelming. We visited the famous beach of Varadero and enjoyed long walks on the white sand. We lived in the former house of the rum mogul, Bacardi; the adjacent villa was reserved for Soviet astronauts. We cruised on an army ship and I had the chance to dive in the warm Caribbean ocean where one could see in the clear water the wrecks of old ships lying at the bottom. I saw a shark leaving a wreck and how skilfully it was shot by a marine officer. The sailors harpooned fishes and caught crustaceans of various kinds. The fresh seafood was delicious, and the large pink shells still decorate our living room.

We made another trip by plane to Isla de Pinos. I knew that this was the Treasure Island of the famous book by Robert Louis Stevenson. I had read this book when I was a boy, and my hair had stood on end. I did not imagine in my wildest dreams that I would ever have the chance to visit the actual setting for "Fifteen men and a dead man's chest, yo ho and a bottle of rum ...". Now the island is called Isla de la Juventud, and there were numerous schools. We visited one of them, attended by hundreds of children of different ages from Mozambique. On the island there was also the prison where the former dictator, Batista, had incarcerated Castro and his friends: the positions of their beds from those days were marked by their photographs, and in addition to Castro, I recognized the faces of some of the other Cuban leaders I had recently met.

We also met Castro's brother, Ramon, and his wife. Ramon was an exuberant, light-hearted renaissance man. His specialty was agriculture and he told us with tremendous enthusiasm about Cuban achievements in this field. We visited a huge cattle research institute, but little seemed to be in progress. Cattle breeding and dairy farming were important activities on the island, but the institute looked grossly oversized to me. I was asked what kind of interferon research should be carried out at this institute, but I could not come up with any good suggestions.

One night we had a party in our villa. Besides my interferon friends, quite a number of others came, including Castro's personal physician, Professor Eugenio Selman whom I had already met. This active, alert man was known to the Cuban interferon workers as "the lively one". He had a large family, ten children as I recall, and followed Castro everywhere. He seemed to have much influence, even in matters related to interferon. Surprisingly, Castro himself then appeared at the party. We had a couple of rum drinks while he smoked cigars (I understand he later gave up smoking). He gave us a number of personal gifts, including the pen rack of black marble which is still on my desk, and a beautiful vase which adorns our living room.

What was the reason for all this Cuban interest in interferon? Clearly Castro himself was behind it. I noticed that all the necessary resources were readily found even in such a small and poor country if the leader with his almost unlimited power regarded the matter as important. Interferon was such a matter, and as the result a magnificent institute was built on a valuable site with amazing speed, and in spite of great difficulties, equipped with top quality apparatus. Because of the U.S. embargo, many of the instruments had to be acquired through intermediary agents which greatly increased their cost. Nevertheless, everything was accomplished quickly and smoothly, since this was the wish of the leader. The best qualified experts were transferred from their jobs to work on interferon research, and young and bright scientists began to be educated and trained in this field.

My impression is that Castro's interest in interferon was sparked off by the visit to Havana of a famous U.S. doctor, Professor R. Lee Clark. I had met this remarkable man when he came to my laboratory in 1979, and he was an impressive and charismatic person. He was appointed physician in chief and principal surgeon at the M.D. Anderson Hospital in Texas in 1946, and during the next 30 years, he brought the hospital to its present internationally renowned position. He was greatly respected in the U.S for his promotion of cancer research and advances in treatment, and he had been advisor to President Nixon on these topics. He came to Finland soon after my article "Why is interferon not in clinical use today?" appeared, and apparently it made a big impression

on him — he called it monumental. His visit to Havana coincided
with the peak period of media "hype" about interferon. Because
Lee Clark knew my work and was familiar with the results of
Gutterman's team, I think he was responsible for transmitting
enthusiasm to Castro.

Many people have wondered why a country as small and poor
as Cuba has invested so much in interferon. I see the situation
in the following way. During Castro's regime, the country has
remained on par with many other developing countries in most
spheres, but there has been much emphasis on the development
of education and health care, and great progress has been made
in these fields. I think these programmes formed as it were a
pyramid, but at the top there was still a place left. A major
research project was needed to form the crown of the pyramid,
one that would benefit the country directly, and would bring
not only honor and prestige but also much needed foreign
currency. For this purpose, interferon seemed in many ways to
be ideal.

The work in the laboratory in Cuba went well, and its staff
made a considerable success with the production and purification
of leukocyte interferon. They also boldly entered the up to date
and fast moving field of gene technology.

In 1983, I made a second visit to attend the First Cuban Seminar
on Interferon, which was held in the large and impressive
convention center. Just before the opening ceremony, I was
informed that I was to be elected the honorary president of the
congress and that I would be expected to give an address. At
this very short notice, I tried to collect in my mind some thoughts
about the development of interferon research and its future
prospects, and about my familiarity with the interferon programme
in Cuba. There were several important national figures sitting
on the stage in front of the big auditorium. When my honorary
presidency was announced, I shook Castro's hand and made a

short speech which was printed in full on the next day in the local newspaper, Granma.

This First Cuban Seminar on Interferon launched a series of similar international meetings in Cuba in which the subject expanded from interferon to cover biotechnology in general. The Cuban interferon workers founded a society called Sociedad Ibero-latinoamericana para Investigaciones sobre Interferon, and invited me to be an honorary member. In 1984 the society started a new publication, Interferon y Biotecnologia; later, the name was changed to Biotecnologia Applicada. I have been a member of the Editorial Board of this journal from the beginning, though I have only played a rather passive role.

I must now give some background to explain the story which follows. In 1981, I was awarded the Erik K. Fernström Grand Nordic Prize, which gave me a considerable sum of money. With the aid of this, Aila and I began to plan to realise the dream of every Finn, to build a lakeside summer cottage. We acquired a piece of land in an out-of-the-way location in Mäntyharju, the place where I was born. I bought a chainsaw, and Aila and I spent our whole summer vacation in 1985 clearing the site. The forest was very thick, and the work was hard but enjoyable. The foundations of the cottage were laid in the fall, and in the spring two local builders started the actual construction. The cottage was supposed to be finished before midsummer when our vacation began, but, unfortunately, the work was still not completed. Nevertheless, we began to live in the half-finished house and on Midsummer's eve, we purchased mattresses for our beds and carried them onto the top of our car to the front of the cottage. The sun was shining, and work was in full swing inside and out. This was the state of affairs, when quite unexpectedly, a black Mercedes-Benz drove into the yard, and to my complete amazement, out stepped Sinikka Hirvonen, my co-worker, together with Impi Haapio, who had for a long time been a housemaid with my parents, two unknown ladies and the chauffeur. Gradually it became clear that one of the strangers was the new Cuban ambassador and the other was her interpreter. Aila invited everyone in for fruit juice and coffee, and, little by little the purpose of the visit became clear.

The ambassador had received a personal invitation to me from Fidel Castro to come to the inauguration of a new research institute in Havana in about two weeks' time. This had put the poor lady to a great deal of trouble. She telephoned my laboratory only to be told that I had just left for my summer vacation, so she got in touch with the director of our institute and said it was absolutely essential that she should talk to me in person wherever I might be and as soon as possible. The director asked Sinikka to act as a guide, and the group headed for Mäntyharju which is about 200 kilometers from Helsinki. Sinikka knew that I was somewhere in that area. She telephoned my parents at their farm at Kivelä, and Impi joined the group to guide them for the last 20 kilometers to my isolated place in the woods.

Now the ambassador put all her diplomatic skills to work to try to persuade me to accept the invitation. However, I did not give in: there was so much enjoyable work to do on the cottage and the best weeks of the summer were ahead. The ambassador was deeply disappointed, but I tried to make amends by promising to send a personal cable to Castro. This I duly did, thanking him for the invitation, and giving reasons for my non-acceptance. I congratulated him on the new institute, and emphasized the importance of research in the field of biotechnology.

One final reminiscence from this ambassadorial visit to our cottage still comes to my mind. After the long ride, the coffee and fruit juice prompted the ladies to satisfy natural needs, and the ambassador inquired which was the way to our toilet. However, as in most Finnish cottages, we did not have a water-closet and our latrine was still unfinished. I had just brought the appropriate plastic part for this in Mäntyharju, and I hurriedly put it in place. Thereupon the ambassador, as the first to patronise it, "launched" our latrine.

On a later visit to Cuba, I saw the new institute and was astonished by its huge size. Manuel Limonta had become its director, and Pedro Lopez, another member of the original delegation who had visited my laboratory, was now in charge of the "old" interferon institute. The interferon seed had grown into a large biotechnology tree.

On one further occasion, I was sent a personal invitation from Fidel Castro. This time the occasion was the 30th anniversary of his revolution, which would be celebrated in Santiago in Eastern Cuba. I understand that Aila and I were the only persons from Finland to be invited: for some reason, Finland would be sending no official government representatives. Nevertheless, being a scientist and not a politician, I decided not to attend the celebrations.

In recent years, I have been saddened to learn of the increasing impoverishment of Cuba and its isolation from the rest of the world. What fate lies ahead for my friends in that faraway country?

33

Why Is Interferon Not In Clinical Use Today?

In 1979, the publishing company, Academic Press launched a new book series called Interferon. The driving force and editor-in-chief was Ion Gresser in Paris; Edward De Maeyer in France, Michel Revel in Israel, Jan Vilcek in New York and myself in Helsinki were the other editors. After thorough discussion, we had collectively formed a plan for the series. This was that there would be a new volume each year which would contain between 5 and 7 articles covering different areas in interferon research. The authors would be invited to write an essay describing advances in their particular field, highlighting current problems and discussing prospects for the future. They would be encouraged to speculate boldly, to put forward hypotheses, and to provide readers with stimulating and provocative insights into their chosen field of research. We hoped that the volumes would reach a wide circle of readers, and would appeal to them by having a format and informal style rather different from those in conventional scientific articles and reviews. The series continued until 1987 and a total of eight volumes appeared.

For the first volume, I wrote an essay which had the same title as this chapter. I had a clear purpose and objective in writing my essay, and this filled my thoughts for several months. At that time, the greatest problem in interferon research was, at least in my opinion, the shortage of interferon. I wanted to emphasize this and stimulate others to remedy the state of affairs.

In 1978, Aila, the boys and I spent a part of my summer vacation on the island of Gotland in Sweden. As we cycled along the beautiful level roads, thoughts about interferon passed through my mind and every once in a while, I stopped to put on paper the ideas that had occurred to me.

As I worked on my essay, many things that filled me with annoyance came to mind. How little time and effort had been devoted to the production of interferon. How much interferon research had been carried out with preparations of miserable purity. How sad that great and wealthy nations and their pharmaceutical companies had been mere passive spectators of the interferon scene, giving no positive and active support for its development. How stupid it was to wait for still more evidence that interferon did have value as a drug when there was not enough interferon available to permit definitive clinical studies. How irrelevant it was to make theoretical calculations of the amounts of interferon required to treat a human being. How stupid and shortsighted to complain about the tremendous cost of interferon, when its future as a medicine would surely depend not on its price but on its value in medicine: if the value was sufficiently great, price would certainly not be a barrier to its use. How utterly wrong were the statements, so often repeated, that it would be almost impossible to prepare sufficient amounts of human interferon for clinical trials: man had already solved far more difficult technical problems such as sending astronauts to the moon and bringing them back alive. I reflected on the seemingly almost unlimited mental and material resources available for the military machine in all countries; if interferon had had some possible use in the event of biological warfare, and so been given the necessary priority, all problems related to its production would certainly have been quickly solved (to avoid any misunderstanding, I must stress at once that I am against warfare of any kind). I thought again of how we had developed production methodology and made this freely available to others, but it had been utilized in only a few laboratories. How vain and theoretical had been the wrangling about which sort of interferon is best, when the only way to find out was to make all the possible kinds of interferons and compare their performance in the clinic.

The essay gave me the opportunity to express in a forthright and frank way all the thoughts which had gnawed away at me for many years. I also included in the article a number of bold predictions about the future for interferon. Today, enough time has elapsed for me to see how wrong I was in many of these. For example, my speculation that there might be a special epithelial interferon has turned out to be completely erroneous. Probably my worst mistake was to grossly underestimate the possibilities of gene technology, even though I had then already started to collaborate with Charles Weissmann in this field. But as a result, I predicted that in the 1980's, most clinical trials with interferon would make use of product derived from human cells, whereas in fact most was made in cultured coliform bacteria. Although I was completely wrong here, I must in self-defence say that this was partly because I was eager to stress the value of the production methods already available when I wrote my essay.

Nevertheless, even today, I am proud of some of my predictions. I saw clearly that the continuing speculation and argument about the superiority of one type of interferon over another was stupid and naive: I predicted that all types would be needed, because each has its specific qualities. I also suggested that the main clinical use of interferons in the future would be in combination with other methods of treatment, and this is certainly the trend today. I also foresaw that long-term treatment with interferon might evoke or even worsen auto-immune diseases, in which the natural immune defence mechanisms in a patient's body lead to the destruction of some tissues or organs. It has indeed been found that interferon treatment may increase auto-immune damage in the liver and the thyroid gland.

My optimistic guess was that, by the year 2000, various interferons would be used in the routine treatment of many diseases, and this will certainly be the case. I also predicted that appropriate interferons would then be widely used for veterinary diseases in animals ranging from mink to race horses, but I now doubt that this will happen for a considerable number of years.

In my own opinion, this essay is one of the most important bits of writing that I have ever produced. It was created with much difficulty but with much pleasure, and I dedicate it to my wife Aila. It has been she who has made it possible for me to devote so much of my energy to interferon, and she has brought into my life things still far more important.

34

Namalwa

From the time I first began making interferon in white blood cells, I was fully conscious of the limitations of the process. Because it makes use of fresh human blood as the starting point, the amount that can be made will always be limited. Blood is not collected merely in order to make interferon; indeed the white cells are obtained as a by-product of the normal procedures of a blood service.

It is true that I made theoretical calculations for the amount that could be prepared in an emergency situation in which leukocyte interferon was the only available method for controlling an epidemic of an otherwise lethal disease. Under such circumstances, it would be justified to collect blood merely for the manufacture of interferon. Even then, the greatest amount that could in practice be obtained would soon be reached. At the start of our studies, I would not have believed what very large amounts of leukocyte interferon could in fact be produced 10 to 20 years later, and this has been a most welcome surprise. Nevertheless, the number of available cells was — and will always be — the main factor limiting the amount of interferon made.

We did an enormous amount of work to obtain as much interferon as possible in our production system. Once induced, white blood cells form interferon rapidly, and after they have made this, they are still alive and in good condition. We tried every trick we could think of to stimulate the cells at this stage to produce interferon for a second time: all were unsuccessful. The cells just would not yield a second crop. There is probably a very good reason for this. From the point of view of a cell and the whole animal, the production of interferon is a key defensive and regulatory event. It is important that production starts quickly, but it is probably equally important that it stops soon so that the amounts of interferon produced are not excessive, for these might have detrimental effects on the cells or animal. This would explain why cells have developed efficient regulatory mechanisms to keep interferon production under control, and why we could not overcome them.

The practical conclusion of all this was clear. For really large scale production of interferon, many more cells will be needed than will ever be available from human blood.

A group of Danish scientists and businessmen paid two visits to my laboratory. They had the grand idea of preparing interferon for use in Man from the white cells in pigs' blood. They worked out how much pig blood should be obtainable from the huge Danish slaughterhouses, and how much interferon this should yield for use in human medicine. I did my best to discourage them. I pointed out that, first, interferon derived from pig white cells should of course be active in pigs, but would have little or no activity in man. They thought it might be possible to modify porcine interferon in such a way that it became highly active in human cells, but this looked to me completely unrealistic. I don't deny that man and pig have much in common, especially as far as behaviour is concerned, but at the level of interferon the two species are quite different. In contrast, the relatively close links between man and monkeys are also reflected at the level of their interferons. My other cause of concern was that even supposing that pig interferon could in some way be made to work in human cells, the human body would in all likelihood recognize it as a foreign protein, and the antibodies consequently formed against it would neutralise its effects.

The Danes worked for some time on this pig interferon project, then quietly gave it up. My own view all along has been that cells from any foreign species will prove of no avail as a source of natural interferon for use in human medicine. For this purpose, interferon must be produced in human cells. But how can huge numbers of human cells be obtained?

With such thoughts at the back of my mind, I read with growing interest in the 1960's reports describing how the white cells in human blood can be used to establish lines of cells that can be repeatedly subcultured. Usually, white cells taken from human blood survive in culture only for a short period of time, a matter of a few days. However, it was reported that in some instances cultures had been established in which the white cells apparently grew and multiplied indefinitely. The first cultures of this type were derived from a malignant tumour called Burkitt's lymphoma. This tumour of lymphoid tissue is rare in almost all parts of the world, but quite common in Uganda and certain other regions of Africa. The tumour derives its name from an English surgeon, Dennis Burkitt, who published the first description in 1958, based on his clinical experiences in Uganda. The tumour occurs most often in children and is usually localized to sites in the facial region; it is highly malignant and soon leads to the patient's death.

The peculiar geographical localization of this tumour, and, later, the discovery in the tumour cells of a new virus, the Epstein-Barr virus, a member of the herpes group of viruses, made the study of Burkitt's lymphoma very popular among scientists in the 1960s. Many laboratories began to work with the so-called lymphoblastoid cells derived from these lymphomas. Among the scientists who became involved in the study of Burkitt tumours were Werner and Gertude Henle in Philadelphia. When they looked at various cell lines derived from Burkitt patients, they noticed that some produced small amounts of interferon apparently spontaneously.

At the Roswell Park Memorial Institute in Buffalo, U.S.A, George Moore had much experience with the culture of lymphoblastoid cells. It turned out that similar "immortal" cell lines could be obtained even from the blood of completely healthy persons. Some of these lines harbored a virus of the herpes group, but in others, no virus could be detected.

I became interested in these cells, and intrigued by their unlimited growth capacity. They did not attach to the glass or plastic surface of their culture container, but formed loose clusters of cells which sank to the bottom. This property made it much easier to grow these cells in bulk. On the other hand, from the point of view of interferon production, there were big problems associated with them. In the first place, it was quite unclear whether these cells could be induced in some way to produce large amounts of interferon. Human blood contains several different kinds of white cells, and we knew that only a small minority of these contributed to the production of interferon in our routine method. Could it be these producer cells constituted the lymphoblastoid cells in the continuous cultures, or another non-producing type of cell?

Another worrying problem stemmed from the potentially malignant nature of all mammalian continuous cell lines. Cells derived from a Burkitt lymphoma were clearly cancer cells, but even the lymphoblasts derived from healthy humans had some similar feature; for example, they had abnormalities in their chromosomes which resembled those seen in cells obtained directly from tumour tissues. During the 1960's, the safety of the different types of cell that could be used for vaccine production was discussed at length. The consensus of opinion at that time was overwhelmingly against the use of a continuous cell line as the source of any vaccine intended for use in man. This was because of fears, which however were only based on conjectures, that such a product derived from malignant cells might cause cancer in the recipient. Was it then worth even considering the production of interferon in lymphoblastoid cells? Yes, in my opinion, it was.

I reasoned that if I could develop a good process for making interferon from lymphoblastoid cells, the resulting interferon could at least be used for work on the structure of human interferon. Also, although the possible risks associated with our leukocyte interferon had been emphasized *ad nauseam* by some individuals and were still being stressed, perhaps all the concern with this and with interferon derived from lymphoblastoid cells was grossly exaggerated. In the final analysis, the essential thing would be to have a purification process that would effectively free the product from all possible viruses. As will be clear, the criticisms directed toward our leukocyte interferon annoyed me considerably, and the unlimited possibilities of lymphoblastoid interferon attracted me. In a way, if I was to come up with a good method for the production of lymphoblastoid interferon, this would form a reply to those criticising our leukocyte interferon.

During the late 1960s, I collaborated closely with Karl Fantes in London. He told me he had obtained a lymphoblastoid cell line from George Moore in Buffalo, and I asked Karl to send me some. The cells of this RPMI 1196 line, which originated from the blood of a healthy person, reached me in January 1969. They grew well in our laboratory, but sadly, produced only negligible amounts of interferon. We tried all the tricks we knew which had enhanced the production of interferon in white cells cultures: nothing worked. At best, the cells produced only a fraction of the amount obtainable from fresh leukocytes. I knew that such cell lines undergo continuous changes during long term culture in the laboratory, and I had myself found that in cultures of another continuous cell line, the HeLa cell, there were sublines which were either good or bad as producers of interferon. I tried infecting RPMI 1196 cells with different viruses; I tried growing them in the continuous presence of interferon, and to select from such cultures sublines of cells that might be good interferon producers. Nothing worked. The yields remained miserably low.

In June 1969, I went to a conference organised by The New York Academy of Sciences to present my data. Karl Fantes had seen in the programme the enigmatic title of my paper "Attempts to prepare interferon in continuous cultures of human leukocytes." When he saw me, he came up at once and asked: "Have you solved the problem?" I had to admit that, unfortunately, my results were most suitable for that as yet unknown periodical, "The Journal of Negative Results." When I came to give my talk, I concluded by saying that, although RPMI 1196 was a poor producer of interferon, it would be worthwhile studying a number of further lymphoblastoid cell lines. This was exactly what I was planning to do.

I had already written to George Moore to ask him if he would be willing to send me some more lymphoblastoid lines, and he replied that he had more than 600 others in his laboratory. He generously promised to send me some. In practical terms, it was arranged that he would send several cell lines in June 1969 to the Waldorf Astoria hotel in New York, the venue for a meeting of the New York Academy of Sciences, and the hotel in which I would stay during the conference. When I saw the extravagant decor in this luxury hotel, I was reminded of a jocular but bitter comment by an American scientist, Joe Pagano, who had complained about the tremendous costs of housing monkeys in some experiments he was conducting; he groaned "it would be cheaper to pay for those darned gibbons to live in Waldorf Astoria than to take care of them in my laboratory."

I brought Moore's cells back to Finland and we began eagerly to work with them. We had problems with the growth of some lines, and they died out, but even the best among the remainder lines were very disappointing, producing very little or no interferon. But I did not yet give up.

My friend, Hans Strander, was very busy in the clinic in Stockholm, but he had nevertheless begun to carry out work with interferon in the laboratory. We planned collaboration and decided to try once more to produce interferon in lymphoblastoid cells. Hans had begun his research career working in the laboratory of George Klein at the Karolinska Institute in Stockholm, and this laboratory was now actively engaged in studies of Burkitt's lymphoma.

Klein had established direct links with surgeons in Africa who sent him small pieces of tumour tissue from numerous Burkitt's lymphomas. He established many cell lines from such tumour tissue, and generously supplied these to other investigators, including his former student, Hans.

Hans and I planned our collaboration in such a way that he would grow the lymphoblastoid cells in his laboratory in Stockholm and induce the cultures with Sendai virus which I would send from Helsinki. He would then send samples from the cultures to my laboratory so that we could measure how much interferon had been made. We proceeded according to this plan, and, in 1972, studied eight cell lines: again, all the results were negative. We still did not abandon our efforts. At the beginning of 1974, we tested 21 new Burkitt cell lines which had been established in Klein's laboratory. At long last, our luck changed. Although two thirds of the lines produced no or very little interferon, there were four that were reasonably good producers. The best of these lines had the name Namalwa, and we decided to concentrate our future studies on it.

Hans sent Namalwa cells to my laboratory, where they soon were growing well. In our studies with them, the results were very gratifying: the same tricks which had improved the production of leukocyte interferon also worked for Namalwa interferon, and we could readily make 20 times more interferon than in the preliminary experiment in Stockholm. With our improved methods, the yield from Namalwa cells was about 20 per cent of the corresponding yield from leukocytes, and it seemed likely that it might be improved still further. But what kind of interferon did Namalwa cells make?

We started to compare Namalwa interferon with leukocyte interferon, and Erik Mogensen joined in these studies. The two turned out to be similar, and clearly different from fibroblast interferon. To our great delight, the purification method which we had worked out for leukocyte interferon could equally well be used to purify Namalwa interferon.

In the summer of 1974, we collected our results into a paper which appeared in the following year in the American publication, the *Journal of Clinical Microbiology*. This paper by H. Strander, K.E. Mogensen and K. Cantell was short, but its two pages were packed with condensed information. It concluded with the statement that large amounts of human interferon could be prepared in lymphoblastoid cell cultures, and that such interferon had the desirable characteristics of leukocyte interferon.

I was very pleased with our results, and considered that I had reached my goal on this particular research front. We now returned to the active pursuit of our own leukocyte interferon production system. The Finnish Blood Center had also begun production in 1972, and its output was rapidly increasing. We had no possible means or indeed need to start on the large scale production of Namalwa interferon. My hope was that our paper would stimulate some pharmaceutical company to work on this.

For my part, I got in touch with the biggest Finnish pharmaceutical company, Orion in 1974, and told them about our then still unpublished results. I said I would be pleased to help them if they were interested in making the Namalwa interferon. The scientists in Orion were indeed interested and they began to cultivate Namalwa cells and to carry out preliminary studies on producing interferon with them. Apparently, however, the technical difficulties, the costs, and the uncertainty concerning ultimate success proved too much for the company's management, and quite soon, the work was abandoned. I was naturally disappointed, but I felt I had done all I could to give a Finnish company the opportunity to join in the development of interferon.

Nevertheless, although our findings did not take root in my native country, things began to move swiftly in England. There, soon after the discovery of interferon in their laboratories, the Medical Research Council had involved three big British pharmaceutical companies in a collaboration aimed at developing it as a drug. They were Glaxo Laboratories, Imperial Chemical Industries (ICI) and the Wellcome Foundation. By 1971, the first two companies had lost confidence in any future for interferon, and had stopped further work. For reasons unconnected with interferon, three scientists who had worked actively on interferon as part of the collaboration left these companies and moved to join Wellcome, where a little work on interferon was still in progress.

Although I was a friend and had been a collaborator with both Fantes and Finter, they were not able to tell me what plans Wellcome had for interferon. Secrecy is a fact of life in industry, but I found this strange and alien to my own philosophy, for I have always spoken openly about my experiments and their results, and about my plans for the future. Be that as it may, I have learned since that in 1974, Norman Finter persuaded the management in Wellcome to explore continuous cell lines as a possible source for the mass production of human interferon. At that time, it was a daring decision, for it was almost unthinkable that a commercial company should even contemplate using cancer cells to make a medicinal product for use in man. I admired greatly his courage and open-mind: this kind of attitude changes the world. It seems it was needed also in the industry of my own small country.

Wellcome first tested various continuous cell lines such as HeLa, but in January 1975, our paper on Namalwa cells brought them welcome news. Karl Fantes soon wrote to me, asking for Namalwa cells, and I passed on his request to our original supplier, George Klein. In all, Wellcome collected some 200 different continuous cell lines, mostly lymphoblastoid cell lines, from different laboratories around the world and compared their capacity to make interferon when stimulated with the particular strain of Sendai virus which we used in leukocyte interferon production. When Namalwa cells turned out still to be the best of all those tested, Wellcome decided to proceed with the manufacture of Namalwa cell interferon on a large scale. By 1978, this interferon was being produced in a 1000 litre capacity tank, and two years later, in tanks that were ten times bigger. With modification of the conditions, interferon yields were eventually obtained that surpassed by a factor of ten those reported in our original paper. The problem of how to purify the Namalwa interferon still remained a matter of crucial importance. Fantes developed a method which purified the Namalwa interferon about 100 000 times, and in so doing, efficiently removed Sendai and other viruses, as well as proteins and nucleic acids derived from the cells and serum. His multi-stage purification process has not been described in detail, but it involves the use of ethanol, as in our own purification process.

It was Finter's idea to add various known bacteria, viruses and nucleic acids to the crude Namalwa interferon and so to provide direct proof of how efficiently they were all destroyed or eliminated during the purification process. The purity of the final preparation was a matter of decisive importance. The approach impressed and finally convinced the control authorities in various countries, initially the United Kingdom, then the U.S.A. and Japan, who had previously been nervous about the safety of the lymphoblastoid interferon product, which was now called Wellferon. On the basis of the purification work and animal experiments, and when all the required extensive laboratory safety tests and checks had been completed, these authorities gave permission for its initial use in carefully controlled studies in man. On the basis of these and many other later studies, and with continued reassurance about the safety and efficacy of Wellferon, Wellcome's lymphoblastoid cell interferon eventually became a registered drug.

Namalwa was a young Ugandan girl with that name who developed a Burkitt's lymphoma and died because of it, in spite of surgery and drug treatment. Her life as an individual thus came to an end prematurely, but the cells born in her body still live in laboratories around the world. I have often looked at photographs of African children with Burkitt's lymphoma in the medical textbook edited by Dennis Burkitt. They show grave little faces which have been distorted in a hideous way by huge tumours. The distress and agony of Namalwa, Akuba, Daudi, Maku, Raji and numerous other children, as well as that of their relatives haunt and distress the reader. What senseless injustice it seems. The story of Namalwa interferon cannot diminish the shock at the fate of this poor child, but I get at least some relief from the knowledge that her disease has helped to alleviate the suffering of many people. Personal ambition is the strongest motive for a scientist, but however hardened he may be, the suffering and agony of a child must move him in whatever quarter of the world he works. Namalwa's interferon only became a medicine as the result of the contributions of many scientists in many parts of the world.

35

A Phone Call From Zurich

One day in March 1978, I was sitting in my office in National Public Health Institute writing a letter, when as so often happened at that time, my work was interrupted by the ringing of my telephone. The call came from a man who introduced himself as Charles Weissmann, a name then unknown to me. He told me he worked in the Institute of Molecular Biology of the University of Zurich, and that he was very interested in interferon, and particularly my own research. Without further ado, he invited me to give a seminar in his institute as soon as possible.

Now I am not in general fond of travel, and at that time, I had a particularly full programme. I was then receiving invitations from very many different quarters, and visitors were literally swarming to my laboratory. I replied right away that I could not possibly make a trip to Zurich in the near future. Weissmann was not discouraged by my negative and perhaps slightly irritated attitude. He outlined with broad strokes the scheme he had in mind for a collaboration, aimed at isolating the gene responsible for the formation of interferon from our human white cell cultures. This sounded exciting even though utopian, so I expressed cautious interest. Thereupon, Weissmann asked whether perhaps he could come for a quick visit to Helsinki to meet me and to tell more about his plans. I was accustomed to a flow of visitors, and my reply was: "Why not? When would you like to come"? "The day after tomorrow", was the answer. His eagerness and spirit made me laugh aloud. I glanced at my calendar and said: "Welcome!"

Two days later, we sat face to face in my office. Two hours passed quickly, and we continued our talks in the cafeteria and then again in my office. His grandiose idea was to isolate human interferon genes and to use them to produce interferon in bacteria.

The idea was not completely new to me. In the previous year, I had read an article by an English scientist, Derek Burke, in the magazine, *Scientific American.* Burke, a former collaborator of Isaacs, had written that it might one day be possible to transfer the gene of human interferon into bacteria, and so make them into "small factories" for the mass production of human interferon. I had thought then that this was a truly fascinating possibility, but one so futuristic that it would not be realised for many years. However, I had been so busy in recent years occupying myself with my own activities, that I had only been able to follow in a very superficial way the startling strides being made in molecular biology.

Charles Weissmann was obviously a bright scientist and I was greatly impressed by his personality. As he explained in a very clear way just what he envisaged, his ideas which had previously seemed to me something from the realms of science fiction began more and more to sound like the basis for a realisable project. I learned only later that Charles was a very well known and very highly respected scientist in his own field. He had worked for years in the U.S. and currently was a Professor at the University of Zurich and Director of the Institute of Molecular Biology.

During the previous summer, Weissmann had met an interferon scientist named Peter Lengyel. I also knew this bright and pleasant research worker, who was born in Hungary but had moved permanently to the United States. Peter had told Weissmann about interferon, and the two men had shared the dream of isolating the gene for human interferon, and using this to make interferon in limitless quantities; if realised, this would open completely new dimensions in interferon research. Peter had described a new method which could be used to detect interferon messenger RNA — this is the piece of ribonucleic acid which carries information copied from the DNA-containing interferon gene in the nucleus of a cell to the "factory" sites in the cell, the ribosomes; here, the coded information carried by the messenger RNA is accepted as an instruction to make interferon protein. In an extremely sensitive newly developed method, the interferon messenger RNA from a mammalian cell was directly injected under the microscope into frog eggs. As a result, the ribosomes in the egg cells made the mammalian interferon in accordance with the message, and excreted the interferon into the surrounding fluid, where its biological activity could be measured in the usual way. The two scientists discussed how they might be able to make use of this sensitive method in order to pick up the mouse gene responsible for forming interferon. The idea was based on the fact that a gene, which is formed of DNA, and the corresponding messenger RNA molecule made from it recognise and bind to one another with amazing precision and efficiency.

Research into other fields in molecular biology in previous years had disclosed a variety of other technical tricks which could be applied to nucleic acids. It had been found that, based on the chemical composition of a particular messenger RNA, the corresponding DNA could be prepared, either as a single strand, or in the natural double-stranded form. A strand of DNA could be cut at will at pre-defined sites in the molecule into pieces, and these could later be joined together again. A portion of DNA could be multiplied by joining the fragment to a particle called a plasmid; this could then introduce it into a bacterial cell where numerous copies would be quickly made.

The goal that Lengyel and Weissmann envisaged was to isolate and multiply, or in the jargon, to clone the mouse interferon gene. The first step would be to stimulate mouse cells to make interferon, and to isolate the mouse interferon messenger RNA molecules formed in them. The next step seemed difficult: the mouse cells would at the same time also be making messenger RNA molecules for many other proteins. How could the relatively few messenger molecules that were sought be picked up from among the myriads of those that were irrelevant?

The solution to this problem would be to convert all the messenger RNA molecules into their corresponding DNA molecules, and then to increase their number in bacteria in the way described by means of plasmids. All these reproduced DNA molecules would then be attached to a suitable solid surface, in the hope that among them, there would be some representing mouse interferon genes. To show this, the surface would be treated with a proven interferon messenger RNA preparation, which would bind specifically and tightly only to these. The surface would be thoroughly washed to remove all the unbound RNA molecules, and then the bound messenger RNA detached and injected into frog's eggs: if these now made interferon, it would prove that the interferon gene really had been attached to the surface.

All this must sound terribly complicated to a reader who is not familiar with this field of science, and it sounded so to me also when it was explained to me for the first time, and even though I was so to speak a semi-professional.

Lengyel and Weissmann decided to start a collaboration along these lines. Lengyel therefore isolated and partially purified mouse messenger RNA, and in the autumn of 1977, sent the first batch from his laboratory in Yale University in New Haven to Zurich, where Weissmann's group started a hunt for the interferon gene. However, during the next few months, a number of events had a major impact on Weissmann's plans, and led him to discontinue his collaboration with Lengyel and instead seek one with me. Together with some ten leading scientists from Europe and the U.S.A., Weissmann founded a company called Biogen, which was intended to provide a channel for exploiting commercially the inventions and discoveries made in the various laboratories. A project to clone the human interferon gene soon became one of the main goals of the new company: success in this endeavour seemed to offer the best prospect of making money as a result of the patents, reagents, and hopefully the new drug which would result. Initially, all the interferon studies were carried out in Weissmann's laboratory, but later another founder-member of Biogen, the Belgian scientist Walter Fiers, also became involved. The scientific board of Biogen decided to support the mouse interferon work already in progress in Weissmann's laboratory, but the highest priority was given to the cloning of human interferon.

No one in Weissman's laboratory had any experience in how to make and assay interferon, and Lengyel's activities were entirely focused on mouse interferon. So Weissmann needed to find another collaborator with as much knowledge and experience as possible in relation to human interferon. From reading the scientific literature, he soon came to the conclusion that my laboratory would form the ideal partner for his project. At the time when he first telephoned me, interest in interferon was mounting steeply, and as the methodology and the equipment needed had already been available for some two years, he was fully prepared to learn that I was already collaborating with some other group in an attempt to clone the human interferon gene.

This then was the background to Weissmann's phone call to Helsinki, but I only understood the situation clearly much later. During his visit to my laboratory, Weissmann said nothing about Biogen or the commercial implications behind the collaboration, and these dawned on me little by little much later. I was interested in the collaboration only from the scientific point of view, in line with my own previous studies on the production of interferon, and again did not think about making money from it. Nevertheless, no doubt, I would have collaborated with Weissmann in just the same way, even if I had known from the very beginning of his connections with Biogen. I considered his chance of success to be very remote, but the importance of the project was obvious. There is a saying in Finland: "Salmon is such a tasty fish that it is worth trying to catch one even if in the end you have no success."

Weissmann and I decided to begin our collaboration immediately. Peter Curtis, a British scientist from his laboratory, soon visited Helsinki, bringing the various reagents needed. We set up white cell cultures in the routine way and induced them to form interferon with Sendai virus. We collected samples from the cultures at hourly intervals, broke up the cells and extracted their nucleic acids by treatment with phenol. The mixture was allowed to form layers in a separation funnel, and we then collected the nucleic acids in the supernatant from the heavier phenol layer. These nucleic acids were precipitated with alcohol, and taken by Curtis to Zurich. There, their messenger RNA was purified about tenfold, and each sample was injected into at least ten frogs' eggs; these eggs were kept in culture overnight. The supernatants were then harvested,frozen and sent to my laboratory in Helsinki to test whether any of the eggs had produced interferon. To our delight, we found there were indeed measurable amounts of interferon, so the system did work. We also soon showed in this way that the amount of interferon messenger RNA in the white cells reached its peak about 5 hours after they were induced with Sendai virus.

Now the problem was to find the needle in this haystack. We were looking for interferon messenger RNA molecules which formed only a minute fraction of all the messenger molecules present in our white cell extracts. Weissmann harnessed the best resources in his big laboratory to the task.

I had most contact with two very bright, skilful and hardworking Japanese students, Tada Taniguchi and Shige Nagata. Tada was the next visitor, bringing with him many reagents. He, Sinikka Hirvonen and I laboured for a long day and night during which we extracted RNA from a hundred billion white cells which had been stimulated five hours earlier to produce interferon. The work involved handling certain reagents in large volumes. This caused a number of problems, and those with phenol were particularly unpleasant for this poisonous and evil-smelling substance caused headache and irritation of the eyes, and readily burnt the skin and one's clothes. Working together gave me the chance to get acquainted with Tada, and I formed a very good impression of this self-assured and self-motivated young scientist: it has been no surprise to me that he has had a very successful career since as an independent scientist. He also made a big impression on Sinikka by presenting her with a fine bouquet of white roses on the following day.

Quite soon after this visit and much to Weissmann's regret, Tada left Zurich and returned to Japan. There he began to clone the interferon gene from human fibroblasts, applying the methodology which he had learned in Zurich: the apprentice became a competitor.

The need for messenger RNA increased. Next, Charles Weissmann himself came to Helsinki, and again we worked until late at night. Sinikka and I learned from him the tricks of how to extract RNA, and from then on we did the work on our own. We became more skilful and obtained more suitable equipment, but nevertheless we still suffered every time from the unpleasant effects of phenol.

In Zurich, Weissmann's group worked hard. The partly purified messenger RNA was converted to DNA, and this was joined to plasmids, and in this way, taken into bacterial cells and copied innumerable times. The progeny of 10 000 such bacteria were individually selected for study. This huge task was carried out by Shige Nagata and by Weissmann's godson, Michel Streuli, who had come from the United States to work in his godfather's laboratory. The hope was that some of these bacterial cultures would now harbour the interferon gene, or at the very least, a fragment of it. However, it would have been an impossible task to test such a large number of samples directly for the presence of this DNA by the laborious method needed.

To reduce the burden, Weissmann had the idea of making pools of 512 bacterial cultures, and looking for the interferon gene in 12 of these pools. From each, the plasmid DNA was isolated and fixed on small filters. These were treated with interferon messenger RNA isolated in Helsinki and purified in Zurich. The RNA that bound to the filter was detached, purified, concentrated and injected into frogs' eggs, and the culture fluids were shipped to Helsinki to be tested for any content of interferon. If there should be any, this would mean that at least one of the 512 progeny bacteria in that particular pool contained the whole or a part of the interferon gene.

Among the twelve pools we tested, eight were consistently negative, but four seemed to give positive results. Unfortunately, however, the interferon levels were very low and, worse still, when we repeated the assays, the results did not agree well with those of the original tests. In spite of the discrepancies, Weissmann decided to place some reliance on the weakly positive results, and further work was therefore now concentrated on the four questionably positive pools. From the 512 individual clones which were the constituents of each of these, eight smaller pools, each now representing only 64 clones, were prepared. These were tested by the laborious method described above. As soon as I had the results of the interferon assays, I phoned Zurich where Charles and Shige waited eagerly: the news was that three of the pools of 64 clones now gave positive results. With one, the results were the most clear cut, and from its components, eight smaller pools of eight clones were made: now three gave positive results.

Immediately the first positive pool was identified, Weissmann started to analyse its components. Among them, he found a clone which repeatedly gave positive results, but occasionally also negative results. It contained a piece of human DNA that was about one-third the length of the interferon gene. Was this really a part of the eagerly sought interferon gene? The DNA fragment was made radioactive and used as a probe to hook out similar but longer pieces of DNA from the positive bacterial cultures. At last, a piece of DNA long enough to represent the full length of the interferon gene was found. The next step was to prove convincingly that this was indeed the interferon gene: could it lead to the production of biologically active human interferon in bacterial cells?

Six months after I started my collaboration with Weissmann, Bob Friedman, an interferon scientist from the U.S., invited me to take part in another similar one. I declined, a decision which clearly pleased Charles. As the end of 1979 approached, we heard rumours that studies like ours were in progress in several other leading laboratories. In September 1979, Tada Taniguchi, now back in Japan, told his former master, Charles, that he had succeeded in isolating the interferon gene from human fibroblasts. It proved later that he was right, but at first this was not clear: he had not then shown that his isolated gene gave rise to biologically active interferon. As a result of these items of news and rumours, the pace of work in Weissmann's laboratory accelerated. When I made my next visit to Charles, I noticed a sleeping bag in his office, and that told me a lot.

At the end of 1979, Charles decided that he would also establish the interferon assay method in his own laboratory, and from then on he was no longer totally dependent on assay results from my laboratory. Samples were sent to my laboratory primarily for confirmation of his own results.

Work was now concentrated on the clone of coliform bacteria containing the piece of human DNA suspected to be the full length interferon gene. These bacteria were grown in quantity and then disrupted so that any human interferon that had been produced would be released into the medium — these bacteria have no mechanism for excreting a protein into the surrounding medium.

The first interferon assays on such bacterial extracts were carried out in Zurich just before Christmas 1979. As usual, interferon activity was measured by testing whether the sample could protect human cells against the destructive effects of infection with a selected virus. The first results appeared totally negative, but on closer examination, there was a hint of a slight protective effect. Charles was just about to leave for Christmas skiing holiday with his family, but before he left, he asked his godson, Michel Streuli, to repeat the interferon assays, but to test the bacterial extracts at higher concentrations. Being a Jew, Christmas was not a religious holiday for Charles, and work in his laboratory continued in full swing. On Christmas Eve, he called the laboratory from a phone booth high up in the Alps and inquired about the assay results. He was fully prepared to hear that these bacterial extracts had given negative results in these repeat assay, but Shige Nagata had truly amazing news: several extracts had given strong protection against infection with the challenge virus! That was the end of the Christmas holiday for Charles. He returned immediately to the laboratory, and progress was made at a frenetic pace during the subsequent weeks and months.

Samples were also sent to my laboratory, and we confirmed the Zurich assay results. Both our laboratory and theirs also showed that the interferon activity present in the bacterial extracts was blocked by specific antibodies directed against leukocyte interferon, but not by antibodies against fibroblast interferon. A number of other confirmatory tests were also carried out in Weissmann's laboratory with convincing results: the coliform bacteria had indeed produced biologically active human white cell interferon. The almost unbelievable goal had been reached.

The paper describing this finding appeared in print in March 1980 in the world-renowned scientific journal, *Nature*. It had eight authors from Zurich, with Shige Nagata as the first and Charles Weissmann as the last, and one author from Helsinki — Kari Cantell. The contribution made by Sinikka was gratefully acknowledged. Before the paper was published, Charles had taken out a patent on the discovery in the name of Biogen. In the previous summer, this company had been on the verge of bankruptcy, but it was saved by agreeing to assign any future interferon patent to an international pharmaceutical company called Schering-Plough in exchange for eight million dollars. This company made a splendid bargain.

I remained blissfully unaware of these financial transactions.

36

I Miss A "Moment Of Decision"

When I was a young boy, I read a book by Stefan Zweig with the title *Moments of decision for Mankind,* which made a deep impression on me. It tells in a masterly way about various fateful moments which had far-reaching consequences on the destiny of the world. Such was the moment that, according to Zweig, decided the fate of Napoleon, and of the world, at the battle of Waterloo. Another was the composition of the Marseillaise by an unknown composer, "a genius of one night", in a flash of inspiration during the tempest of the French revolution. Yet another was the failure to close one small gate due to some almost inconceivable act of negligence, that led to the fall of Constantinople.

I regard the discovery of interferon in 1957 as another decisive moment for mankind, but thinking about such moments, I shall always remember one associated with my collaboration with Charles Weissmann.

At the end of 1978, the search for the interferon gene was in progress in the very laborious and tedious way which I have described above. The gene could only be identified indirectly through its ability to bind interferon messenger RNA, which could then in turn be measured. To pick out the gene from within the ten thousand bacterial clones was a truly formidable task.

The strategy adopted was to test a few large pools, and then pools of fewer and fewer of their constituents, and finally the progeny of a single bacterium, but all the way, there were numerous hidden problems and snags that threatened to abort further progress. Even this extremely laborious set of procedures ultimately identified only one single positive bacterial clone. By the end of the year, the prospect of success was still completely uncertain.

This was the time when Charles decided to try a short cut on the basis that bacteria containing a full length copy of the interferon gene would indeed produce biologically active interferon. In the end, this expectation was confirmed, but at the time, there were two worrying facts that cast considerable doubt on the likelihood that this approach would succeed. First, human interferon is promptly secreted into the medium surrounding the cell in which it is made, and such a protein destined for secretion is made in a precursor form. At this stage, it carries an extra piece, a so-called signal protein, which directs it to a special site in the cell; there the signal protein is removed and the protein is excreted. As coliform bacteria are unable to secrete proteins, it would be necessary to disrupt their cells and look for any interferon formed within them. However, it seemed very possible that the interferon precursor protein would contain the signal protein and might in consequence not be biologically active, i.e., might not inhibit virus multiplication (it only turned out later that the presence of the signal protein does not in fact inhibit the activity of interferon).

A second concern stemmed from the belief at that time that interferon molecules contain sugar; these would not be present in interferon made within coliform bacteria, for these bacteria do not add sugars to the proteins which they make. Thus if sugars were indeed present and essential for the biological activity of mammalian interferon, this so-called recombinant interferon from bacteria would be inactive (later, it was shown that sugars are indeed present in some of the leukocyte interferons, but are not essential for their biologically activity).

Towards the end of 1978, one hundred pools each of 64 bacterial clones were prepared in Charles' laboratory. The bacteria in the pools were disrupted, the cell debris was removed by centrifugation and the supernatants were sent to me for assay. The hope was that at least one of these bacterial pools would contain a whole interferon gene and produce enough active interferon for us to measure.

We tested the extracts by the plaque assay method I have already mentioned. In this procedure, suitable human cells are grown in flat culture dishes until they form a continuous sheet of cells on the surface. A series of dilutions is made of each test material, and each dilution is added to one or more culture dishes which are incubated overnight to allow time for any interferon present to act on the cells. Then, all the dishes are infected with a virus: if this grows in the cells, it destroys them. The areas of resulting cell destruction are called plaques, and can be seen with the naked eye and counted. If there is sufficient interferon in the dilution tested, this will protect the cells so that no plaques or fewer plaques are formed. Thus the plaque count shows whether or not the sample contains interferon, and if it does, how much.

Counting plaques is dull, boring work, but has been part of my weekly routine for 30 years. The method yields accurate results but because it is so laborious, it has been abandoned in many laboratories in favour of other methods for assaying interferon. I still use it because it has certain advantages, and while counting plaques for many hours is undoubtedly tedious, the results obtained have often given me moments of great delight.

From the point of view of assay, Weissmann's bacterial extracts were particularly unattractive. They were toxic to the cells, and partially or totally destroyed the cell layer in the culture dishes. Some extracts nevertheless clearly did inhibit virus multiplication, but I could not get reproducible results, and the toxic damage to the cells made the results difficult to interpret. If I tested the samples at greater dilution, the toxicity was less but their virus-inhibitory effects were also reduced or lost altogether. I told Charles on the phone that I did not trust these weakly positive and non-reproducible results.

At that time, there were many things in my personal life and in the laboratory which seemed to me more important and interesting than assaying these troublesome bacterial extracts, and also I just could not believe they would contain interferon. Charles too was quite ready to abandon his short-cut approach as apparently it did not yield clear-cut results. With hindsight, however, it is certain that some of the bacterial extracts which we studied in the fall of 1978 definitely did contain interferon. If pools had been prepared of fewer bacteria rather than of 64, the interferon content would have been higher and the results more certain. If I had had more confidence in the success of the project, I would have tried harder to confirm the weakly positive results: I could have used a more sensitive version of our routine assay, and proved, by using specific anti-interferon antibodies, that the activity detected was in fact due to interferon.

What would have been the consequences? The interferon gene would have been cloned a year earlier than actually happened, and biosynthetic interferons would have become sooner available for use in medicine. Thus I had in my grasp the key to bringing forward by one year the fruition of a scientific achievement of great benefit to mankind. Such an opportunity comes seldom, and it is also seldom that a scientist later realises how close he has been to an important discovery. Be that as it may, I know that in the fall of 1978, I missed a "Moment of decision."

37

The Value Of A Scientific Observation

There are some similarities between achievements in art and science, but there is a crucial difference. It is self-evident that if Sibelius had not composed his violin concerto, or Shakespeare written "Hamlet", these masterpieces would never have existed. But it is just as certain that even if Watson and Crick had not discovered the double helix, the structure of DNA would have clarified before very long by others. Equally, if Isaacs and Lindenmann had not discovered interferon, other scientists would have found it — indeed, probably sooner rather than later. I believe that no discovery in the field of natural sciences is unique in the same way as the creation of a piece of art. Scientific phenomena are there already in nature, and just await discovery. All scientific advances could and would have been made by others, if they not been made already by the scientists to whom they are credited. So from my point of view, the true standard for the value of any scientific discovery is time, the unit being the number of months, years or decades gained for mankind by the discovery. If it had not been for the actual discovery, how long would it have been before that information came to light? I am not sure that this concept of a "time standard" for discoveries is entirely an original idea of mine, but I certainly do not recall seeing it anywhere bluntly stated in this way.

Of course, the weakness of my reasoning is that one can never know what would have been the exact course of events if things had developed differently. How would science have advanced if there had not been for example Newton, Darwin, Pasteur, Einstein, or Fleming? What was their contribution in terms of "time saved"? We can only speculate. Nevertheless, in this connection, the following example is interesting.

Some years ago, two young American scientists, David Baltimore and Howard Temin, independently discovered a new enzyme of key importance, called reverse transcriptase. This happened so simultaneously that their findings were described in two papers published in 1970 in the very same issue of *Nature*. From the point of view of these two scientists, the timing was perfect. Had one written his paper even two months later, he would have missed the Nobel prize, which, in the event, both scientists shared. It can be argued that, from the point of view of the benefit to mankind, this duplicated effort was totally unnecessary, for either one of the two publications would alone have provided the important new information. Nevertheless, independent, simultaneous discoveries are not all that uncommon: when the time is ripe and there is sufficient background information, there are many minds ready to take the next step. So a great scientist is not irreplaceable in the same sense as a great artist. All around the world, teams of scientists work away busily, often focusing their efforts on a small problem. It is the persevering efforts of many such teams that push forward the frontiers of knowledge a tiny bit at a time. Only occasionally is there someone with the originality, ability and courage to take his own path, and this individual stands out clearly from the common ruck. Nevertheless, even though scientists may not be irreplaceable in the same sense as are artists, even a mediocre scientist can be invaluable through his personal influence on his co-workers. Also, a work of only modest intrinsic scientific value may, when published, kindle a new idea in a fellow scientist and thus, indirectly, lead to an important finding.

In an article dealing with the nature of research, I wrote in 1980 "The works of most contemporary composers are almost without exception rather dull and uninteresting, dispensible pieces, lacking the striking originality which make for example, Schubert's Winterreise, Sibelius's violin concerto and Kosma's Les feuilles mortes supreme and imperishable achievements in their own spheres." On the day after this appeared in print, one of our leading composers telephoned me in a state of great annoyance. He had been deeply hurt by my evaluation of modern works, and during our long conversation, he stressed the point that a composition may be despised today but highly appreciated a century later. I did not argue, although, in my own opinion, the likelihood seemed infinitesimally small. A "misunderstood" artist or scientist can, of course, always seek consolation from a thought like this, but I believe this is mere self-deception; the fact is that the names and contributions of extremely few scientists or artists from our times will be remembered after a hundred years.

The scientific community is not always ready to accept important new information when it first becomes available. Gregor Mendel's experiments with beans, published in 1865, brought to light the basic rules of heredity, but the importance of his observations was only recognised after more than three decades. Similarly, the significance of the tumour virus found by Peyton Rous in 1911 and of the jumping genes described by Barbara McClintock in the 1940s were not understood and appreciated for many years. These are well-known examples of remarkable findings which had to wait a long time before they received their proper recognition. Perhaps at the time, each was too revolutionary or too far outside the scope of what was popularly held.

I should also point out that a new piece of information can lead to important applications, but can also lead the scientific community on to a false trail. This is exemplified by the events which I described in the chapter "The shadow of Maurice Hilleman": in the late 1960s, interferon scientists became so over-persuaded of the value of a new class of chemical inducers of interferon that they lost interest in methods for making interferons. Many years were lost before it was realised that interferons are actually better drugs than at least these inducers. Unfortunately, scientists too often behave like animals in a herd rather than as independent individuals, and this behaviour is aggravated by the feverish competition for research funds.

Today, the ever-increasing flood of information is a real problem for every scientist. Computers give unparalleled help in the search for items of some particular interest among all those in the enormous information jungle, and indeed this is now a profitable business area. However, among their positive features, computers have introduced a strange new phenomenon into the world of science. From their huge data bases, they can count how often a particular scientific paper is referenced in other scientific publications. Such "citation indexes" have increasingly come to be used in the evaluation of scientific work. The presumption is that the more often a paper is cited, the more valuable it must be, and equally, the more often the work of a scientist is cited, the better he must be. "Citation" lists prepared for individual research workers are increasingly used as criteria when filling a post or awarding a grant or prize. In my opinion, such a list can be very misleading, as this example from my own experience shows.

The original paper by Alick Isaacs and Jean Lindenmann describing their discovery of interferon, and a book *The Interferon System* by Bill Stewart which was published 22 years later, have both amassed about the same number of citations. Does this mean that the scientific value of the two publications is the same? Stewart's book has been a very useful reference source for many scientists because it contains a vast amount of information which is presented in a clear and logical fashion, but it contains hardly any new information. In my judgement, the scientific value of the original discovery by Isaacs and Lindenmann is many hundreds of times greater.

38

New Information About Interferons

In the memoirs of the leading nuclear physicist, Andrei Zacharov, there is a chapter in which he describes his research in some detail. His field is not too easy for a layman to understand, and at the beginning of this chapter, Zacharov advises a reader who does not find the topic interesting just to glance through it, or even to skip the chapter altogether. Following his example, I will say that those who find this chapter boring or difficult can skip it without losing the tenor of my story.

In the early years after its discovery in 1957, interferon was thought to be a single protein which could inhibit the multiplication of many different viruses, but had no other biological effects. This view of interferon has changed radically in the course of the many years of research since. It has gradually become clear that not all interferons, not even all human interferons, are identical. For example, those produced by white blood cells and by fibroblasts are similar in many ways, but clearly different, and even more different is the so-called "immune" interferon which Fred Wheelock discovered in 1965. I am ashamed to say I did not understand the importance of this last type of interferon for many years, and have only relatively recently become aware of its unique and interesting role.

When at long last in the 1980s, precise information became available about the different forms of human interferon as a result of applying gene technology methods, it became evident that human cells harbour nearly a score of functional genes, each responsible for the formation of a particular interferon protein. These genes and their interferon products are now classified according to the Greek alphabet: the former "leukocyte" interferon has become alpha interferon and there are 13 different variants, or "subtypes" of this; "fibroblast" interferon is now called beta interferon; among the interferons induced in white blood cells by Sendai virus, there is an omega interferon which has some resemblance to both alpha and beta interferons; and the old "immune" interferon is now called gamma interferon — in its structure, as in its properties, it is markedly different from the other interferons.

The mechanisms responsible for the regulation of these interferon genes have become the subject of intensive research, and this has also shed light on the general regulatory mechanisms controlling all genes. When I read about work in this area, I remember my late friend, Tapani Vainio, who had already thought about such studies in the 1960s. Had he lived, I wonder what he would be studying today.

Much is now known about the structure of interferons. Like other proteins, they are made up of a chain of amino acids, usually 166, which forms a long coiled chain. The sequence of the particular amino acids in each of the different interferons is known, and the three-dimensional structure of the molecules has been worked out in considerable detail. Some interferons do have sugar molecules attached to their amino acid chain, but their role is still rather unclear.

Interferon molecules are quickly excreted from the cells in which they are made, and they act on cells in the neighbourhood. To do this, they must first bind to one of the many specific receptors, i.e., special sites for attachment, on the cell surface. This binding triggers a chain of events that culminates with a signal or signals passing to the cell nucleus. Exactly how this happens is a very active area in current interferon research.

When a cell is stimulated in this way by an interferon, it begins to make tens of different proteins, which in turn are responsible for the actual inhibition of viral replication, the slowing of cell division, and the numerous other effects produced by interferon treatment. However, no function has yet been found for most of these proteins induced by interferon.

Interferons play an important role in the defence of the body against viral infections. In their turn, many viruses have their own weapons in their fight against interferons. Knowledge concerning the tactics adopted by different viruses in this warfare is increasing rapidly.

Initially, interferons were studied almost entirely by virologists, but after some years, immunologists, i.e., scientists who study the body's defence mechanisms, also "discovered" them, and interferons now form a topic under very active investigation in their field. Interferons are also involved in defence against certain bacterial infections. Nevertheless, it seems probable that nature developed interferons not primarily as weapons against microbes but because they have other functions of more fundamental importance, but not yet understood in detail. Interferons seem to be important regulators of cell growth and differentiation. It used to be assumed that cells do not make interferons under "normal" conditions, that is to say unless specifically stimulated in some way. However, by using modern, very sensitive methods, small amounts of interferons can be regularly detected in those who are in good health.

How do interferons inhibit tumour growth? Numerous possible mechanisms have been found. Interferons slow the growth of cancer cells; they force them to differentiate into a more mature and less malignant form; they reduce the growth of new blood capillaries into tumour tissue, which consequently dies because of the lack of nutrients; they alter the tumour cell surface so that it is more readily recognised and attacked by the body's defence forces; they make white cells more aggressive in their attack on cancer cells; they regulate the production of certain hormones which in turn can affect the growth of the tumours. Which of all these actions is important in a particular tumour is not yet clear.

In the early years of interferon research, I felt that I was familiar with the entire scientific literature dealing with interferon. This comfortable feeling did not last long. Over the years, knowledge about interferons has increased so tremendously, and has expanded and ramified into so many new directions that no one can any more command more than a small fraction of the whole. This problem is common to all scientists, for new information accumulates at such a rate that it is truly difficult to keep up with it. If one wants to master a particular field really thoroughly, one must be prepared to narrow the one's field of interest. This can be dangerous, for it is essential also to keep one's eyes on the new information accumulating in biology in general. It is not easy to decide how much time to devote to reading and how much to experimental research, and I have no satisfactory answer to this problem.

Since the discovery of interferon, a large number of other proteins have been found of about the same size, and resembling interferons in a general way in their production and actions. These are now collectively known as cytokines. Many are produced by different members of the white blood cell family. Many were discovered by chance and often simultaneously and independently in different laboratories with the result that at one time, one particular cytokine might have been called by several different names; it only gradually became clear that all these names, in fact, referred to one and the same substance.

Cytokines were at first produced in minute quantities in cell cultures, and their effects were measured by laborious biological methods. However the whole field of cytokine research has been dramatically changed thanks to modern methods of gene technology. The ill-defined and troublesome "factors" of a few years ago have turned into molecules of known composition which can be produced in large quantities. As a result, several cytokines have already been entered into clinical trials, and many others will follow suit. They are being tried primarily as treatments for infectious diseases and tumours, but some seem likely to be found useful in other target diseases.

Many cytokines are growth factors, and for example, some control the growth and differentiation of various blood cells. They also regulate the functions of the cells forming the body's defense forces. The cytokines form a very heterogeneous group, ranging from those with a number of different actions, like the interferons, to those which are highly specific and exert a particular effect only on a single type of cell.

There is a complex network of interactions between different cytokines and between cytokines and interferons. Some cytokines enhance and others inhibit the actions of interferons, or the actions of other cytokines on interferons. Both cytokines and interferons carry messages from one cell to another, and in this respect, they resemble hormones. However, in general, hormones circulate in the blood and carry their messages over long distances from one part of the body to another. Cytokines and interferons, on the other hand, carry their messages over a very short distance to target cells nearby. They can even regulate the functions of the cells from which they are derived.

For someone who has long been involved with interferons, it has been exciting to follow developments on the cytokine front.

These have in many aspects mirrored research on interferons, which to a large extent have been the forerunners in this whole field. In their development over the years, the story of many cytokines repeats the history of interferon research. However because of new methods, research into cytokines has progressed very much faster than in the early years of interferon.

For a long time, interferons were believed to be unique, but we now know that they belong to the large and growing family of the cytokines. In one respect, though, they are still unique: their common characteristic is that they can inhibit the multiplication of many different viruses and in an effective way. No other cytokine has exactly these properties.

Advances in research will continue to open fields beyond our present vision in the world of interferons and cytokines. There are certainly exciting new discoveries awaiting explorers.

39

Mathilde, Once Again!

Without doubt, Mathilde Krim must have been greatly heartened by all the publicity which interferon received after her very successful congress in 1975. She made plans for a new and even larger meeting to be held four years later, which appropriately enough is the interval between successive Olympic games. This time, organisation was a lot easier, as she obtained financial support from some twenty sponsors including the NCI, the NIAID, the American Cancer Society and many pharmaceutical companies and Foundations.

Mathilde also established an interferon laboratory in the Memorial Sloan-Kettering Cancer Institute and gathered around her a stimulating group of research workers, including Bill Stewart. Bill had changed both his wife and laboratory, moving from De Somer's laboratory in Belgium to that of Ion Gresser in Paris. Meanwhile during visits to Poland, he married a young girl, Marzenna Wiranowska. Due to his influence, this charming girl started to work on interferons, and she moved with Bill to the United States and to Krim's new laboratory in New York. Once there, Bill started a collaboration with me aimed at identifying any sugars present in my leukocyte interferon.

Another colourful character who joined Mathilde's team was Budd Colby, a tall yankee with Indian blood in his veins. No less of character was a third, Chris Tan, a very alert and quick-witted Chinese from Singapore, with the happy gift of being able to think very fast on his feet. I remember wondering at the time how long a collaboration between such a temperamental group would last; the answer was not very long. After a few years, the group broke up and the interferon laboratory was closed, but many things happened before then.

Bill Stewart led a very hectic life, working long hours in the laboratory during the day and writing a textbook on interferons at night. This was published in 1979 with the title *The Interferon System*, and was an enormous success: it was at the time the most comprehensive and up-to-date treatise on interferons and the timing was perfect. Interferons had become a more and more popular subject and were being studied from many points of view. Stewart's book was used widely as a source of information and as a general guide to published information about interferons, and soon became an indispensable reference book for those working on interferon.

Stewart had an exuberant personality and strong opinions. Interferon workers were quite sharply divided into two camps, those who liked him and those who loathed him. I could see his weaknesses, but nevertheless I liked him. He founded a periodical, the *Journal of Interferon Research*, and an International Society for Interferon Research, both of which have a permanent place in the history of interferon research. He ran both the Journal as its editor, and the Society as its Secretary-General, in a somewhat dictatorial manner, which made many of his opponents extremely angry, and led to unsuccessful attempts to unseat him from his posts. But he handled the Society's international relations very skilfully, and as a result he had more support overseas than in his native United States.

There was a nucleus of active interferon workers who participated in many international meetings and so gradually became more and more acquainted with one another. Some became friends, others enemies — likes and dislikes are as frequent and passionate in the world of science as in the world of arts, and I, like many scientists, have strong sympathies and antipathies. Differences in characters, competition for power and prestige and many other reasons can complicate the personal relationship between scientists working in the same field, the same city, or even the same institute. Sometimes this leads to moments of comedy.

There was some friction between the two distinguished interferon workers in Kyoto, Professors Yoshimi Kawade and Tsunataro Kishida. There was also friction between Charles Chany and Ion Gresser who had both worked for many years in Paris. When Kishida visited Gresser's laboratory, the latter asked his guest with a malicious twinkle in his eye: "How is your friend Kawade?" In a flash, Kishida replied "And how is your friend Chany?"

Krim's second interferon meeting took place in the spring of 1979. Some sessions were held in the Rockefeller University and others in the Memorial Sloan-Kettering Cancer Institute. Among the more than 300 participants there were Nobel laureates and other distinguished scientists as well as influential representatives of industry and the media. Mathilde invited Jean Lindenmann and me to act as honorary chairmen.

The meeting went well and, from my own personal point of view, the atmosphere was much more relaxed than it had been at the meeting four years earlier. The various presentations gave a clear overall picture of the state of the art in interferon research, both in the laboratory and in the clinic. I believe that at last, the importance of overcoming the then chronic shortage of interferon became obvious to everybody. In spite of all its wonderful properties, interferon could do no good to anyone until there was enough available. So the importance attached to the development of methods for making it, and especially to our own method, became greater and greater. At that date, the media had already started to give some coverage to interferon. I recall a taxi journey trip from our hotel to the meeting, during which the talkative driver asked Harri Nevanlinna and me where we came from and where we were going. When we told him that interferon was the subject of the meeting, he got very excited and started asking numerous questions. When at last we stopped in front of the Rockefeller University, he exclaimed sadly: "Boy, I wish I could join you".

40

Interferon Becomes A Superstar

Towards the end of the 1970s, interferon became the subject of an increasing torrent of publicity, as in the eyes of the media, it acquired the status of a wonder drug. It seemed it could be used successfully to treat many viral diseases and above all, cancer — diseases that were common, frightening and difficult to cure. Its "magic" status was further magnified by the fact that it was in such short supply and extremely costly. The illnesses of some particular individuals also played a role. Senator Edward Kennedy, the last survivor among the legendary Kennedy brothers, was a major public figure, and when his son had to have a leg amputated because of a bone cancer, the media quickly realised the connection with interferon: this was the very form of cancer which Dr. Hans Strander was treating with interferon in Sweden.

Interferon also reached the headlines when both the Shah of Persia and the film star, John Wayne, developed forms of cancer. I do not know whether in fact they were ever treated with interferon — certainly, we did not send them any from Finland — but their disease helped to make interferon even more widely known.

After Mathilde's second meeting, the publicity surrounding interferon increased enormously. The science writer, Al Rosenfeld, among others, listened diligently to the lectures and interviewed many of the participants. He then wrote an article about interferons which was published in the July issue of the magazine, *Life*. It was filled with enthusiasm, anticipation and optimism; it also included a photograph of me, and the text described me as some kind of hero.

The story filled me with very mixed feelings. Naturally it tickled my vanity, but at the same time I wondered with apprehension what price would have to be paid when such great expectations were not quickly realised. However, I was sincerely happy about the attention which my country received in consequence. How was it possible, people asked, that a country as small as Finland produced nearly all the interferon used in the whole world. A somewhat derogatory term, " Finlandisation" was in use at that time in leading newspapers and magazines in many countries, denoting undue dependency on the Soviet Union. This term gave a negative picture of my native land, and so our interferon now gave it some welcome positive publicity.

Rosenfeld's article on interferon was also published in the *Reader's Digest*, and in this way obtained circulation throughout the world. In Germany, the magazine, *Stern*, emphasised the role of Finland in interferon research and put special emphasis on interferon's price. It was more expensive than gold and in fact, the most high-priced substance in the world: one gram of interferon would cost 100 million Deutsch marks i.e., as much as 25 electric locomotives!

Articles about interferon were published in many countries. With growing concern I read titles which proclaimed that interferon was a new miracle drug, and ranked it in the same category as penicillin. I foresaw stormy times ahead.

In spring of 1979, a reporter from the American magazine *People* came to Finland to interview me. His article appeared in the July issue and contained photographs of me handling eggs in the laboratory with my technician, and another of me boiling eggs in our kitchen at home with Aila. In another large photograph, which was later copied in several magazines, I peered out from between two large round bottles containing interferon.

When such articles appeared, the inevitable result was that a small avalanche of letters of various sorts landed on my desk. For example, I got a friendly letter from New York from a man by the name Robert Cantell, who was astonished by our common name. He told me this was extremely rare in the United States; his ancestors had come from England in about the year 1800, but had originated in Ireland. I wrote in reply that four generations of my ancestors had been church cantors, and that our family name was derived from the Latin word cantare, meaning to sing. He quickly sent an enthusiastic reply, telling me that he was a singer by profession.

Interferon was featured increasingly in the media. Newspapers and magazines such as *Time, Life, Stern, L'Express, Mainichi Shimbun* and *Reader's Digest* published comprehensive reports dealing with it. Already in 1970, the BBC had put out a programme on interferon in their television series, "Horizon" which featured my laboratory. Ten years later, the BBC made a second programme about interferon, lasting 50 minutes, for the same well known series. This contained film and interviews carried out in my laboratory and in the Finnish Blood Center. The programme was shown all over the world and received much attention.

The famous U.S. news anchor-man, Walter Cronkite, came to Finland to interview me for his television series "The Universe". He finished the interview with a surprising question: "Do you believe in God?" I was not willing, however, to give an account of my relation to God in this programme, which had nothing to do with matters of faith.

All this publicity for interferon was both harmful and useful. The worst thing was the completely unrealistic hope which it raised in patients and their families. Because the potential targets for interferon therapy seemed to include many of the most feared and most refractory diseases known to man, there were numerous people drowning in despair and anxious to grasp this apparent life buoy. For many, the wonderful new drug seemed to offer the last ray of hope.

Those working with interferons viewed the situation with very mixed feelings. Of course, the attention and publicity received by interferon was more than welcome, and that directed at the research workers themselves was flattering. Nevertheless, we were all filled with growing concern; we realised all too well that interferon could not possibly live up to the grossly over-exaggerated expectations which had became focused on it. The animal experiments and studies carried out in patients with interferon had indeed shown some promise, but they did not justify overwhelming optimism. I saw clearly that there would be a high price to pay for all the publicity, and that the disappointments certain to follow would provoke strong counter-reactions. I could see very well what great annoyance was being given to doctors caring for cancer patients; they were bombarded with requests for interferon from patients and their relatives, and must have had to spend much fruitless time dealing with these, and at a time when there was so little reliable information about the true efficacy of interferon, and virtually none was available to try.

It was only natural, too, that the tremendous publicity directed at interferons greatly irritated scientists working on other subjects, and this was to some extent aggravated when the attention focused on interferon led to the diversion of research funds to this topic at the expense of others. I believe that many were genuinely and maliciously delighted when all the publicity was eventually followed by the inevitable wave of disappointment and scorn.

Who should be blamed for all this excessive publicity? It reflected, as I have said, great interest of the general public and dread about cancer, and their ardent hope that research would lead to some new and effective treatment. The media played fully on these fears and hopes and in so doing generated the huge pressures, hopes and expectations that became focused on interferon. Personally, I came to understand bit by bit what a vast industry the world of communication is, and how fiercely competitive it is.

Those in the new and expanding biotechnology industry also took clever advantage of the high profile of interferon at this time. Biogen, the company set up by Weissmann, soon encountered a tough competitor in the U.S., Genentech. Scientists there succeeded in cloning interferon genes only a few months later than those in Weissmann's team, and then there was a feverish race to be the first to exploit it commercially. Several other companies soon joined in; these were mostly small and operating with venture capital, but taking advantage of the publicity surrounding interferon, they turned themselves into public companies and their share price soared on the stock exchanges. Soon, knowledgeable and long-term entrepreneurs were joined by short-term speculators whose aim was to make a quick profit by taking advantage of fluctuating share prices and exchange rates. Interferon became an instrument used to manipulate the science policies of nations, and by individuals in their struggle for power.

Overall, no one single person or factor can be given the blame for all the excess of publicity that became directed on interferon, or for the resulting gross over-optimism that was generated. Many people must share the responsibility. I have since attempted to appraise my own behaviour during this period of "hype". Did I for my part seem to provide a basis for the excessive optimism? When I re-read and recall all the interviews I gave, I have not found these to contain gross exaggerations, which I now regret. I tried hard to emphasise how sketchy clinical knowledge and experience at that time was, and especially to stress how little interferon was available. Nevertheless, although I feel quite sure that what I said in public during these mad years were for the most part realistic, perhaps I should have been more outspoken in criticising the grossly over-optimistic and quite unrealistic pictures that some others painted.

To me, it was extremely frustrating that no effective use was made in the U.S.A. of the production methodology which we had developed. Nevertheless, it was of course of great benefit to the Finnish Blood Center, even if not for the progress of interferon research worldwide, that for 10 years, all the interferon used in the U.S. was purchased from Finland.

Biogen and Genentech both soon established close ties with large pharmaceutical companies, and these joined in the development of the interferon genes which each had cloned. In only just over a year, the first interferons derived from bacteria were entered into clinical trials. These studies began to yield results which did not live up to the sensational expectations, and very soon, the pendulum swung to the other extreme. Public attitudes and medical opinions towards interferon rapidly changed to become extremely critical or frankly scornful and its reputation melted away when it now appeared to be "just a lot of hot air". Such conclusions were just as hasty and unjustified as the earlier excessive expectations.

From my personal point of view, the changed situation was not pleasant: having been one of those regarded as a key figure in the field of interferon research, I was in a way held responsible for all the disappointment. Nevertheless, I felt that the new clinical studies with interferon were going rather well, even if slowly. New information was accumulating, and although many of the clinical trials results were negative, here and there, there were positive findings. I did not see any steep jumps or falls in the potential that interferon offered; these were just optical illusions created by the media. When eventually positive clinical results again rose above the threshold of media interest, there was talk about the rebirth of interferons. I knew that, in fact, development had been on a more or less continuous ascent, even though the slope seemed gentle.

From other points of view, all the publicity was helpful to interferon research. For example, the wife of a Texan oil billionaire, Leon Davis, read the article about interferon in *Time* magazine in 1979, and told her husband about it. He became so interested in the subject that he collected 10 million dollars from his friends to set up an Interferon Foundation. This Foundation provided funds that enabled Jordan Gutterman at the M.D. Anderson Hospital in Houston to purchase interferon from the Finnish Blood Center for clinical trials. Also, the great public interest in interferon undoubtedly made Biogen, Schering, Genentech, Roche and many other companies more willing to invest in its development. When the bubble seemed to burst, some of these companies no doubt came to regret their interest. But by then they had invested so much resource in interferon that it became difficult to abandon further development. In the long run, the pharmaceutical industry played a crucial role in developing interferon as a drug. We live in a world of market economy, and no new drug has a future without the strong backing of some pharmaceutical protagonist.

At the beginning of the 1980s, I thought that overall, all the publicity had done more harm to the development of interferon than good. In perspective, I now see things in a more positive light, and ultimately, the publicity was beneficial. I say this even though from my personal point of view, it hit me badly during the most critical period in the development of interferon.

41

Who Will Get Interferon?

During the whole of the 1970s, the lion's share of all production of interferon worldwide took place in Helsinki. The Blood Center steadily increased its output, until finally, it used the leukocytes from nearly 100 000 voluntary Finnish blood donors each year for its production. Thousands of litres of crude interferon were produced, and gave rise to several hundred thousands of megaunits (a megaunit is a million units) of purified interferon. These were used both for laboratory experiments and for studies in patients.

All the financial aspects of making interferon were handled by the Finnish Red Cross Blood Transfusion Service, which received considerable sums of money from its production. The purchasers included, amongst others, the American Cancer Society, the National Institutes of Health, the Interferon Foundation in Houston, and the Karolinska Hospital in Stockholm. The Blood Center in turn supported my own research by paying the salaries of some of my technicians. This was an excellent arrangement from my point of view, especially as in consequence I did not have to spend time writing applications for grants to support my research.

As the volume of interferon made in the Blood Center increased, I was able to reduce that in my laboratories, and I focused our activities even more on various aspects of interferon research.

I also had considerable amounts of interferon available and complete freedom to use them for whatever purpose I wanted.

As the publicity about interferon increased, there were more and more requests for supplies. These came from scientists, practising physicians, and patients and their relatives, and from practically all parts of the world.

There were many scientists who wanted potent human interferon for experimental purposes, and I sent them all they requested, provided the amount was reasonable. Naturally, I provided this without charge. I calculated later that we sent over 1500 packages of interferon to nearly 450 scientists in 40 countries. My generosity was usually gratefully acknowledged but was sometimes also indirectly criticised. Investigators in Germany said that the easy and free availability of potent human interferon from Finland played a part in delaying the production and purification of various animal interferons. It did not seem worthwhile to embark on the production of e.g., rat interferon when human leukocyte interferon, which has some modest activity in rat cells, could be obtained free of charge from Finland.

While it was easy to take a positive attitude toward almost all requests for laboratory use, I had many problems with requests for supplies for clinical use, most of which came from abroad. I tried to evaluate the numerous proposals thoroughly and to discuss them in detail with the investigators. All through, I stuck firmly to one principle: the interferon prepared in my laboratory was not for sale.

Hans and I continued our discussions about the use of interferon for cancer treatment. We knew of course that there are hundreds of different forms of cancer, and interferon had so far been tried for the treatment of only a few. Perhaps some particular types of human malignant disease would prove very responsive to interferon therapy. Therefore, it was not sensible to put all one's eggs in one basket; instead, the limited supplies of available interferon should be used for careful studies in several different forms of cancer. The problem was, however, that if only a single patient or even a few were treated, the results would only reveal quite dramatically positive effects of interferon. To pick up lesser but nevertheless possibly very useful positive clinical effects, extensive studies would be needed, which would take much time to complete and consume large amounts of interferon.

Many clinical trials are so-called multicentre studies. This means that investigators in several hospitals participate in the same study, so that all patients needed to give a conclusive result are entered sooner, and the answer to the problem investigation is obtained more quickly. I was surprised to find how difficult it was in practice to organise such collaborations. The problems stemmed mainly from such defects in human character as envy, hunger for power, ambition and clannishness. The morals of laboratory scientist are certainly no better, but a doctor in clinical practice obviously has a more direct influence on the fate of sick people. Among the most bitter disappointments in all my studies with interferon was the failure of attempts to get certain Finnish hospitals to join in some multicentre studies which I considered very important.

When I now recall the many studies for which I supplied interferon, I have to admit that much was wasted by being used in bad studies which did not yield any reliable information. As a laboratory worker, I only began to realise gradually all the problems involved in clinical studies. I wrote in an article once that my own laboratory held the world record for wasting interferon. This may still be true.

Perhaps my biggest problems came when individual patients or their relatives wanted supplies of interferon. Especially towards the end of the late 1970s and in the early 1980s, letters poured in to me from all over the world. They came directly to my laboratory or to my home, or came indirectly through legations, embassies and various intermediaries.

When interferon was the cover story in *Time* magazine and the BBC televised its interferon programme, there was a sharp increase in the number of requests. Many were very moving, sometimes enclosing a photograph of the patient. The letters were full of distress, anxiety, suffering and despair. At first, I tried to reply personally to every letter, but in time this became impossible, and I had to answer most requests by sending a copy of the "standard" reply which I compiled. This explained to the person making the request that the clinical use of interferon was at an experimental stage, that we had only very limited supplies, and that these were being used in carefully planned and controlled studies so as to obtain reliable information about the potential beneficial effects of interferon as soon as possible. In the meanwhile, unfortunately, we could not supply interferon for the treatment of individual patients.

Requests which came by post could be dealt with in this way, but those which came by phone took up much time. They came from all quarters of the world and at all times of the day and night. For a while I wondered whether to change our home telephone to a secret number, but finally decided against this. On the telephone, I tried again to explain the situation as objectively and sympathetically as I could. Nevertheless, I remember one phone call from Paris which came in the small hours of the night in which a man urgently demanded interferon for the treatment of his wife's brain tumour. In spite of my best intentions, I could not keep myself calm and to the point, and I treated the desperate man quite brusquely for disturbing me at such an inappropriate time.

To my astonishment, many people came to my laboratory even from very far away to beg for a supply of interferon. Such unexpected visits caused me much embarrassment for they not only took up my time but meant that my visitors had fruitlessly expended their own time and money.

I recall vividly a group of three men from Naples who were very hard to get rid off. They tried to appeal to my feelings by showing photographs of a little girl with cancer, and they tried to buy interferon, increasing their offer to astronomically high figures. Finally their behaviour became overtly threatening. I was left with the near certain feeling that the men represented the mafia. I placed in the same category a man from the U.S. who bombarded me in the laboratory and at home with phone calls and letters.

We heard from a number of different sources that interferon was for sale on the black market at incredibly high prices. Some visitors brought me samples of interferon they had obtained in this way, and we found these contained very little interferon, or even none that could be detected. It was depressing and disgusting to see how some people were prepared to take advantage of the distress of others in this way. We never found out for certain what the sources of these "interferons" were, but there were clues that seemed to lead to countries in the eastern bloc.

One bitter memory still comes to my mind from those times, although I would much rather forget it. In 1979, a Japanese professor, Tadayoshi Fujita, chief physician of Akura Hospital and professor of Aichi Medical College came to see me. He had come via Stockholm where he had visited Hans Strander, and he showed me his data on the treatment of sarcomas. They looked so impressive and convincing that I gave him some interferon for future studies. The next year my Japanese friend, Professor Kishida, wrote to tell me that Fujita was a swindler, a physician certainly, but not a professor. He had used our interferon to treat a number of patients with various forms of cancers and had charged them large sums of money. I wrote immediately a furious letter to Fujita and demanded an explanation. I emphasised that I had never sold interferon to any patient; I said that it was the moral responsibility of all investigators to use the precious material in a way that would yield the greatest amount of useful clinical information. I never got a satisfactory answer. I had made a grave error of judgement, which I long regretted. Kishida told the newspaper, *Mainichi*, about Fujita's actions, and subsequently, one of their reporters spoke to me on the telephone and then wrote an article about the matter.

I was also particularly annoyed because, partly due to my action, interferon acquired for a time a bad image in Japan where I had many friends who were fine interferon scientists such as Professors Kawade, Kishida, Kuwata and Nagano.

We very seldom broke away from our basic principle that we would not provide our interferon for the treatment of any individual patient. One of the exceptions resulted from the visit of Professor Sir Ernst Boris Chain to my laboratory in the spring of 1977. I knew this famous man by name, of course, for he had shared the Nobel prize of 1945 with Alexander Fleming and Howard Florey for the development of penicillin. He did not ask for interferon for himself but for a friend. This was the Agha-Khan, the president of the Iranian oil company, who had a chronic herpes virus infection in his eye and now needed a corneal transplant. Our interferon had already been used with some success in England and Germany for the treatment of herpes infections of the eye, as I shall describe later.

Sir Ernst came to Helsinki and sat in my office, a figure from the pages of the history of medicine. He was a small, kindly, lively man who resembled in appearance another famous Jewish scientist, Albert Einstein. He was suffering badly from low back pain, but came to Finland in person because his request had wider implications. He wanted to use his connections with the Agha-Khan to get Iranian oil in a roundabout way into Israel, or at least this is how I interpreted his somewhat vague explanation. I called Harri Nevanlinna and we promised him the interferon for his friend. Later, I had two friendly letters from him, but

I never heard if our interferon was ever used and if so, with what success.

In December 1979, I received the very sad news that my old friend and co-worker, Kurt Paucker, had cancer. He had had back pain for a long time, and indeed already suffered from this problem when we worked together in Philadelphia. He complained again about it when we met at Mathilde Krim's meeting in New York in 1979, but this time the pain turned out to be due to metastatic deposits of an advanced cancer in his spine. I immediately offered supplies of our interferon for his use. Kurt was quickly enrolled as one of Tom Merigan's patients so that with the approval of the Food and Drug Administration, he could start treatment with interferon without delay. He tolerated this well, and for a time, it did seem to be beneficial. I thought of our joint studies from almost twenty years ago which showed for the first time that interferon could inhibit cell division, and I hoped from the bottom of my heart that this same property would stop the growth of Kurt's cancer cells. But it did not happen; his cancer progressed and he died in April 1980.

A year later, the international interferon congress in Rotterdam was dedicated to Kurt's memory. During the opening ceremony, I gave with moist eyes the memorial address for my dear friend, who had done so much for interferon research and had meant so much in my personal life.

42

Interferon In The Eye

One of my first studies with interferon was an attempt in 1959 to influence the course of experimental virus infections in the eyes of rabbits by giving drops of interferon. I have already described these studies, and I never forgot that it might be possible to treat similar virus infections of the eye in man with interferon. I was only able to come back to this problem at the beginning of the 1970s when at long last, I had potent human interferon, and I focused my attention on the treatment of infections of the eye with a herpes virus.

After the first infection with a herpes virus, it often finds a way to persist in the nervous tissue. Thereafter for the most of the time, it remains silent, but it may occasionally become reactivated and then produce blisters in the skin, the genital mucous membranes, or the eye, which later become ulcers. In the eye, a herpes virus infection usually appears as a superficial expanding wound on the cornea, and the eye becomes inflamed and very painful. Even without any treatment, the lesion usually heals within a couple of weeks but it often leaves behind a scar on the cornea. As the condition often recurs, sometimes frequently, there is increasing scarring which may eventually lead to impaired vision, and sometimes even to blindness. Such problems with herpes in the eye are relatively common, especially in warmer climates.

Until a few years ago, a herpetic ulcer was treated by surgical removal of the damaged layer of cells from the surface of the eye, but today, the treatment is to give eye drops containing one of the potent specific antiviral drugs now available.

In my original study in rabbits, the rabbit interferon I used gave no protection to the eyes against herpes infection. However at the beginning of the 1970s, we had human interferon which was roughly a 100 000 times more potent than my original rabbit interferon. As a general rule, an interferon derived from the cells of one animal species works well only in the cells of the same species. However, there are some exceptions to this general rule and, by good fortune, one such exception is the human leukocyte interferon which works well also in rabbit cells, and so can be used for experiments in rabbits.

At that time, there were two leading figures in the field of research into herpes virus infections of the human eye: Professor Herbert Kaufman in the United States and Professor Barrie Jones in England. There was considerable competition between the two, and they were not the best of friends. Both became interested in interferon, and so got in touch with me, as only my laboratory had potent human interferon available in sufficient amounts for their planned studies. The two adopted different strategies. Kaufman did not offer collaboration; he knew exactly what he wanted and purchased the amount of interferon he needed from the Finnish Blood Centre. In my opinion, his research plan was poor, but I had no way of influencing it: he proposed to dilute the interferon so much that I felt quite sure his studies would give negative results.

Jones proceeded differently: he contacted the British Scientific Committee on Interferon, and as this Committee had no suitable interferon to offer him, Norman Finter wrote to me on its behalf. He asked if I would be interested in a collaboration with Barrie Jones. Indeed I was. This marked the start of correspondence between Barrie, Norman and me, and then a collaboration which lasted for several years.

The first task was to see if our interferon preparations would protect the eyes of animals against experimental herpes infection. Kaufman employed monkeys and Jones rabbits as their experimental model, and both obtained clearly positive results. Because of our collaboration, Jones had much more potent interferon at his disposal than Kaufman, and so the team in England was better equipped for exploring the relation between the interferon dose and the response to treatment. Their studies led to astonishing findings. Not only was the protection afforded by interferon long-lasting but it was also essentially unaffected by the number of times the interferon eye-drops were applied each day: the outcome was determined solely by the concentration of interferon applied. In this respect, interferon behaves quite differently from synthetic anti-herpes drugs, for these must be instilled into the eye many times a day; a single drop of potent interferon can protect an eye against viruses for 24 hours. Later, Jones showed that these findings from the rabbit experiments were directly applicable to the treatment of herpes infections of the human eye.

Both rivals, Jones and Kaufman, with their teams gave their results in several papers presented at a congress in Stanford, California, in August 1975. The competition between these two powerful individuals gave a great impetus to progress in this area of research.

From my point of view, however, the best feature of the Stanford congress was that I made the acquaintance of two young German research workers, an ophthalmic surgeon, Rainer Sundmacher and a virologist, Dieter Neumann-Haefelin, both from Freiburg. They described a study in which they had used leukocyte interferon prepared by a German pharmaceutical company in herpetic eye infections. Although their results had been negative, I was very impressed by the careful way in which they had carried out their study. I realised at once that we had interferon that was at least a hundred times more potent than that used in the German study, and I was delighted when the two proposed a collaboration with me. This was the beginning of a series of research studies which lasted over ten years. I could see that Rainer would become a star in this field.

The studies proceeded along two main lines: prevention and therapy. For the prophylactic studies, patients were chosen who had already had several recurrent herpes infections in the eye. Half the number of patients put a drop of interferon into the eye each day for several months, while the other half instilled a drop of an inactive mock interferon and acted as controls. Both preparations, interferon and placebo, were dispensed into coded vials of identical appearance so that neither the doctors nor the patients knew who was using which sort of eye-drops. Such a study is called double-blind (as I described in my chapter with this title), and although this sounds ominous when one deals with an eye disease, it is the best and most reliable way of establishing whether a new drug actually works.

Sadly, the results from this study were most disappointing. Kaufman in Florida had already reported negative results obtained in a similar study, but this did not surprise me as I felt sure he had diluted the interferon too much. But I was bitterly disappointed when Sundmacher's study also gave completely negative results in spite of the much more potent interferon used. This completely failed to prevent recurrence of herpes virus infection.

Why did locally applied interferon work so well against experimental herpes infections in the eyes of animals, but failed completely to prevent recurrence of herpes infections in man? The explanation is that in the recurrent human disease, the virus reaches the surface of the eye from within. It gets there by moving up inside a nerve fibre from the site in the central nervous system where it is silently present. The virus then starts to multiply in the cells inside the cornea, but these are not accessible to the interferon molecules which are too large to penetrate so deeply. Interferon can be effective when dropped on to the surface of the eye, but only against a virus which similarly infects through the outer surface. Even today, there is no known way of preventing recurrent herpetic eye infections.

Not only did all the efforts on the prevention front fail, but the early therapy studies looked equally unpromising: the interferon eye drops seemed not to accelerate the rate of healing of herpetic ulcers. However, Rainer Sundmacher and Barrie Jones independently decided to try combining interferon treatment with removal of the infected tissue by gentle scraping. I supplied them both with interferon, and their results were in good agreement. Sundmacher's study was the more extensive and I shall describe it some detail.

In October 1975, I sent to Freiburg coded vials containing interferon eye drops. All looked exactly alike but each vial was given a code letter: vials B, C and D contained interferon, and vials A, E and F placebo. Rainer anaesthetised the eyes of his patients and gently removed the diseased corneal tissue. Each patient then received into the eye two drops each day of one of the coded preparations until healing was complete. Daily samples were taken from the eyes and tested for the presence of virus by Dieter in Freiburg.

At the beginning of May 1976, a total of 40 patients had been treated. We agreed that on a given day, I would send Rainer the code for the vials and he would post the clinical results to me. So our letters crossed each other on the way, as we awaited for them with equal excitement in Freiburg and Helsinki. The study had been carried out critically and carefully in all respects, and now was the decisive moment. Would it put an end to the whole idea of using interferon in the therapy of herpetic eye infections?

To our great delight and relief, the results were excellent. The herpes ulcers in patients who received interferon healed significantly faster than those in the placebo group, and the virus disappeared sooner from their eyes. So the study showed beyond doubt that interferon was clinically useful for the treatment of at least this human virus disease.

Now we set as our goal turning interferon into a practical medication for these eye infections. In Helsinki, we modified the purification method slightly so as to obtain the highest possible amount of interferon in a single eye drop; in Freiburg, Rainer determined what dose and time of administration gave the best results. The studies went along well.

At the end of the 1970s, new and more potent chemical drugs such as trifluorothymidine (TFT) and acyclovir (ACV) became available for the treatment of herpes in the eye. Would interferon potentiate the action of these drugs? I began to look at this question, first in collaboration with Rainer and later with a Dutchman, O.P. van Bijsterveld, and a French scientist, Joseph Colin. All these studies showed convincingly that the combination of interferon and one of these synthetic drugs was clearly more effective than the drug alone.

In the early 1980s, the application of genetic engineering methods started to yield human interferon produced in bacterial cells, as described in the chapter on "A phone call from Zurich." In 1983, Rainer set up a comparison between our natural leukocyte interferon and a biosynthetic interferon from a German pharmaceutical company in the therapy of herpetic eye infections. Both turned out to be effective, though the natural interferon appeared slightly the more effective. The results were published in 1985, and led to registration of the recombinant interferon in Germany and Austria as a agent for treatment — in combination with a drug, TFT or ACV — of herpetic eye infections of man.

I had been involved in the development of interferon for this purpose for a quarter of a century. The work had never received anything like the attention of the cancer studies carried out at the same time, for they went ahead outside the arc light glare of publicity that at one time surrounded the latter. The studies in herpes in the eye had progressed in small but logical steps. I felt that my personal role and the role of Finnish interferon in this field were complete, and I experienced great pleasure and satisfaction to see the brochure for the German interferon drug. At last, something concrete had been achieved as a result of the persistent efforts over many years.

To my disappointment, I have since seen that interferon has in fact been very little used in the treatment of herpes in the eye. This is largely because such good results are also obtained when the newer synthetic drugs are used on their own. Nevertheless, if interferon is combined with them, the results are clearly better still. Why then is this combination not used more generally? I believe the explanation is very simple. The pharmaceutical industry is not interested in this use of interferon, because the market is so small. Only little interferon in total is needed for the complete treatment of a patient, and so this use of interferon will not generate much profit. Money plays a decisive role in the market economy of today!

43

Interferon In The Nose

Acute respiratory infections are among the most common diseases of mankind. The average person contracts several respiratory infections each year. Since the symptoms of a cold are usually mild and short-lived, its importance is often minimised. Yet, in total, colds cause a great deal of trouble and nuisance and have a considerable social impact. They are leading causes of absenteeism from work and school, and their indirect adverse effects are huge, even though hard to measure. How many instances of unsatisfactory performance at work, problems in human relations and how many traffic accidents, for example, should be attributed to the common cold? Also, even a mild infection may be followed by broncho-pneumonia or some other serious complications, and in addition, there are the much more severe respiratory infections such as epidemic influenza which are sometimes fatal.

There are very many different viruses that can produce what a layman calls a cold. Research, especially in England, has provided much valuable information about these viruses and the diseases they cause, and much of this work was carried out in the Medical Research Council's Common Cold Research Unit. This was set up in 1946 in a former American war-time hospital near Salisbury, and there, numerous human volunteers have acted as guinea-pigs in studies in which they were deliberately infected by putting a cold virus into their nose. This made it possible to follow the development of the subsequent cold, and to study factors that influence its course.

Vaccines provide the best and most effective means for preventing viral infections, but this approach seems nearly hopeless for colds, because there are too many agents which may be the cause. Also, there is at present still no effective medication against any one of these agents. *Homo sapiens* has visited the moon and learned how to split atoms, but he cannot yet suppress a common cold.

Knowledge about the various common cold viruses began to accumulate just at the time of the early years of interferon research. In 1960, David Tyrrell, the Director of the Common Cold Research Unit, and a scientist who had worked with Alick Isaacs on interferon, discovered the most important group of the agents causing colds, the rhinoviruses, and showed how they could be grown in culture in the laboratory. Since then, more than a hundred different rhinoviruses have been identified — it is virtually impossible to make a vaccine cocktail of so many. What is needed to prevent and treat colds is an antiviral drug that is broad-spectrum, i.e., is effective against very many different viruses. This is the reason why, soon after interferon was discovered, Isaacs and Tyrrell and their colleagues from the pharmaceutical industry were so anxious to test it against the common cold. Interferon seemed to have just the right properties: its broad spectrum of antiviral activity included effects on respiratory viruses such as the rhinoviruses and influenza viruses, and it seemed devoid of harmful effects when applied to cells in culture. Interferon was discovered in London, and only 90 miles away from the Common Cold unit in Salisbury. Facilities there for studying the prevention of colds were better than anywhere else in the world. The only thing needed before tests could be started was a supply of suitable interferon, and Committee on Interferon set about to try to get this.

For the first study, interferon was used that had been prepared in cultures of monkey cells. It was sprayed into the nose of a group of human volunteers, while others were sprayed with saline. On the following day, all the volunteers were infected with a rhinovirus, given as drops instilled into their nose. The results of this experiment were completely negative. The volunteers in both groups developed colds in exactly the same way, which showed that the monkey interferon preparation had given no protection whatsoever. Probably it contained far too little interferon, but unfortunately, there is no precise information on this point.

Soviet scientists embarked directly on various clinical studies with interferon in the 1960s. They included studies in respiratory virus infections, but these had some surprising, or even quite incomprehensible features. For example, they used chick interferon, and claimed that this gave positive results in the treatment of human diseases. Even at that time, it was generally known that chick interferon has no activity against a virus infection of human cells.

By the end of the decade, human leukocyte interferon made in the Soviet Union was under test for the treatment of a variety of diseases. These preparations were crude and contained very little interferon, but it was claimed that they had activity in several conditions. At the 1967 congress in Siena, my talk was followed by one given by academician V.D. Soloviev from Moscow. He described how intranasal interferon had proved effective in the prevention of influenza during an epidemic. Later, Russian scientific journals published the results of many extensive studies with interferon in the Soviet Union. They all reported favourable results obtained not only in the prevention but also for the treatment of influenza and other respiratory infections. On the basis of such studies, interferon was soon licensed as a drug for sale in Soviet pharmacies.

My attitude toward these results, like that of probably all other interferon workers in the western world, was extremely sceptical from the very beginning. Over the course of years, many visitors brought me interferon which they had purchased in the Soviet Union: we found that this was not only very impure but also contained only a few hundred units of interferon per millilitre. Our crude interferon preparations contained at least a hundred times more, and our concentrated interferon at least a hundred thousand times more interferon than those in the Soviet pharmacies. When glasnost and perestroika finally led to the break up of the Soviet Union, these interferon preparations silently vanished from the shelves of Russian pharmacies. I do not know whether they had ever done any actual harm to the recipients, but they certainly did not enhance the reputation of interferon, or the status of Soviet biomedical science.

In the early 1970s, I supplied the U.K. Committee on Interferon with some of our leukocyte interferon for use at Salisbury. In the first experiment, 180 000 units of interferon were sprayed into the nose of the volunteers, but again, the results were negative. At that time, Tom Merigan decided to spend six months of study leave working in England. He went to Salisbury and put all his energies into a further attempt to inhibit a rhinovirus infection using interferon. In this experiment, his co-workers were David Tyrrell himself, and two other scientists from the Common Cold Unit, Sylvia Reed and Thomas Hall. Interferon was sprayed into the nose of volunteers on a total of 39 occasions, each receiving in total 14 million units, while those in a control group received mock interferon. Neither the scientists nor the volunteers knew who received which. The spraying started one day before the virus infection and continued for three days after. Now at long last, conclusively positive results were obtained. Interferon almost completely inhibited the rhinovirus infection in the treated volunteers. The results were published in the well-respected English medical journal, *The Lancet*, in 1973. There was, however, still one weak point in the study. While it was indisputable that the preparation had conferred protection on the volunteers, a critic could legitimately suggest that the effect was not due to interferon itself, but to some impurity or impurities present in the preparation.

In 1976, George Köhler and Cesar Milstein in Cambridge developed a method by which they could make large amounts of what they termed monoclonal antibodies — later, they were awarded Nobel prizes for this discovery. Monoclonal antibodies can be used for many purposes, including purification of protein, and David Secher working in the same Cambridge laboratory began to try to make monoclonal antibodies against interferon. He immunised mice with Namalwa interferon which Karl Fantes at Wellcome provided, and rats with our leukocyte interferon. Derek Burke was his co-worker in the former study, which proved successful sooner than that in rats. In 1980, Secher and Burke published in *Nature* an article which described the production of monoclonal antibodies against Namalwa interferon. David then began to use the monoclonal antibody for the purification of our interferon. The antibody was bound to a jelly-like substance which was packed in a glass tube column. When an interferon solution was slowly run through this column, the interferon molecules adhered tightly to the bound antibodies as they passed by, whereas molecules of other impurity proteins ran straight through the column. The column was flushed with a salt solution to complete the elimination of these impurities, and then a small amount of an acidic solution was passed through it. The bond between the antibodies and the interferon was broken in the acid environment, and pure, concentrated interferon ran out of the column.

Meanwhile, another bright English scientist, Geoff Scott, had joined the Salisbury laboratory. He was an energetic individual, and amongst other things, piloted himself in a small aeroplane on his way to attend interferon congresses in different parts of Europe. He was the prime mover in a study which compared the efficacy of our usual partially purified interferon with that of the pure interferon prepared by David Secher in preventing rhinovirus infections. The study was performed according to the state of the art in a double-blind fashion, with controls receiving an inert placebo preparation. A total dose of 90 million units of the pure interferon was administered, and the course of the experimental rhinovirus infection was closely monitored clinically and by virological measurements. The result was completely convincing and pleased me enormously: the pure interferon gave almost complete protection against the rhinovirus infection.

Now even the last doubting Thomas had to believe that interferon sprayed into the nose can indeed confer protection against a rhinovirus cold. This looked like a truly significant achievement, but unfortunately, something surprising and disturbing emerged during the course of the study. The interferon caused unpleasant effects such as stuffiness or mild bleeding from the nose in a proportion of the volunteers who were sprayed with it.

I had had great hopes for these studies. I never thought that interferon prepared by our method would become a practical drug against colds, for if it were to be used for such a common ailment, the demand would be truly enormous, and we could not possibly make enough. My aim was merely to use our interferon as a model to show whether interferon might be a suitable remedy for colds. If the results looked promising, the pharmaceutical industry would undoubtedly become interested, and find ways of producing sufficient amounts of interferon by other methods. Indeed, when these other interferons became available in the early 1980s, they were soon tested to see if they could prevent respiratory virus infections. It was found that the results with both recombinant interferons derived from bacteria and lymphoblastoid interferon produced from Namalwa cells, were essentially the same as those with our leukocyte interferon. They protected against infections with a rhinovirus, but had little or no effect against many other respiratory viruses including influenza viruses.

These studies began with volunteers at Salisbury, but were soon extended to the field i.e., to people living a normal life and liable to contract a cold naturally. Again under these conditions, intranasal interferon protected against rhinovirus infections, but unfortunately, the disadvantages of the interferon treatment became even more obvious. Interferon produced an inflammatory reaction in the nose and led to local irritation. Most people tolerated being sprayed a few times, but if the interferon treatment was continued for more than about a week, there were increasing problems. The nose became congested and secreted watery mucus and blood; in fact, the treatment produced symptoms very much like those resulting from the infections which it was intended to prevent. Did interferon perhaps even make things worse?

These results greatly surprised me, and I include them among the few big disappointments in all my studies on interferons. I had looked under the microscope at hundreds of cultures of cells which had been treated with interferon. The cells looked in excellent condition even though they were protected against infection with a wide variety of viruses. I had without any hesitation expected the same results with the cells in the human upper respiratory tract, i.e., by spraying sufficiently potent interferon up the nostrils it was hoped to give the superficial cells in this region protection against many viruses but without causing any adverse effects. Even after I knew that various adverse effects resulted from injected interferon, I was still confident a local application to the respiratory tract cells would be more or less harmless.

Now, with hindsight, I can see how naively optimistic my attitude had been. There is a world of difference between a single sheet of cells growing on the surface of a laboratory bottle and the complex tissue that forms the superficial layer of cells lining the respiratory tract. The latter is part of the body, and white blood cells will be attracted to the cells on which the locally applied interferon has attached; these white cells will be activated and will release a number of cytokines, resulting in a complex chain reaction which leads to the symptoms observed. This at least is my current hypothesis, though I freely admit I do not know exactly why interferon is irritant in the nose.

I am not yet willing to believe that we have reached an impasse in this line of research. If I were young and working in the pharmaceutical industry, I would like to tackle this problem. I would first study individually the 15 different natural human alpha-, beta- and omega interferons; perhaps among all these, there are some which would be tolerated by the human nose. Then it might pay to modify these natural interferons by genetic engineering methods. Perhaps some structural modifications might provide a molecule which would still inhibit viral multiplication but would not be locally irritating. If there is indeed such an interferon, it might be possible to give it at even higher doses and perhaps it would then be found effective even against respiratory viruses other than those of the rhinovirus family. It would also pay to look into other possible methods of applying interferons locally, and techniques which might prevent the nasal irritation. Perhaps the efficacy of the treatment could be increased by the simultaneous administration of some anti-inflammatory drugs. Indeed, there are a number of recent results showing some promise from the combined use of interferon and anti-inflammatory drugs in the therapy of colds.

There is so much that could be done. Even though the problems seem so great, success in this field would more than justify the struggle.

44

The Breakthrough

In the 1970s, our interferon had been used in clinical studies in several countries. The most impressive results were those obtained in the treatment of herpetic eye infections by Rainer Sundmacher in the studies which I have described. However, these seemed of interest only to a small group of specialist eye doctors, and their general significance was not recognised. In the middle of the decade, Tom Merigan and his co-workers at Stanford showed that interferon could affect the course of chronic hepatitis B infections. This modest beginning opened up a huge new line of research, but it was a long road from their preliminary findings until interferon became a routine treatment for the various forms of chronic hepatitis.

Hans Strander noticed in 1971 that warts disappeared in a patient during interferon therapy, and his astute observation had far-fetching importance. It led to the treatment of laryngeal papillomas with interferon, first in Sweden and later in many other countries. Later, the papillomavirus studies were expanded to include wart virus infections of the genital tract.

In malignant disease, treatment with interferon gave several promising results, but none were as impressive as those just mentioned above. The study by Hans of patients with osteosarcoma had a great impact, because it was largely responsible for bringing interferon into the limelight. However, the results did not prove convincingly that interferon was efficacious for the treatment of this particular disease.

Studies by Hans, Tom Merigan, Jordan Gutterman and their teams showed that several solid tumours such as myeloma, lymphoma, renal cancer, breast cancer and melanoma could shrink, and sometimes even disappear during interferon therapy. Nevertheless, such patients who responded were only a small minority of all those treated, and unfortunately, it was impossible to tell in advance which patients would benefit from interferon therapy and which would not.

Especially in relation to cancer treatment, it became more and more obvious that treatment with interferon was unlikely to be sufficient on its own. Undeniably, interferon had favourable effects on some forms of cancer in some patients, but the overall effect was rather weak. This of course was to be expected. Interferons are regulators which affect a variety of normal cells in the body. It was unreasonable to expect that they would destroy advanced tumours selectively or rapidly. To find out how useful this relatively weak anticancer activity would be in practice, it would inevitably be necessary to carry out large, well planned and fully controlled trials. Optimally, treatment with interferon should start as early as possible in the course of the disease, and be combined with other methods of treatment. It would take years or decades for such studies to reach their end and provide answers. All this was very clear to me, but still I cherished a secret hope.

In 1981, there was an international interferon congress in Rotterdam, and Jean Lindenmann and I were invited to give introductory lectures. In mine, I made public the dream which lurked at the back of my mind. It stemmed from the fact that there are at least two hundred different forms of human cancer — indeed, any human cell may be transformed into a tumour cell. Interferon had been tried for the treatment of cancer for only some ten years, during which only very limited amounts had been available. Thus nothing was known about its possible effects in most forms of cancer. It was my fervent hope that among the many still unexplored, there would be one or more cancers that would prove to be as sensitive to interferon as those benign tumours, laryngeal papillomas.

Jordan Gutterman first began to treat patients with Finnish interferon in 1978. This very energetic individual has sometimes been criticised for seeking too much publicity, and for premature publication of preliminary results. It must, however, be added that he and his team have made a number of pioneering discoveries in relation to the clinical use of interferons. Working at the M.D. Anderson Hospital, in Houston, Texas, one of the largest and most respected cancer institutes in the world, he has had access to patients suffering from all the various forms of cancer, and to the most modern available methods for diagnosis and therapy.

There is a particular rare form of leukemia, called hairy cell leukemia, because of the long hair-like threads which protrude from the surface of the malignant cells (it has nothing to do with the hairs of the head or skin). The cancer cells originate from the normal B lymphocytes present in the white cell fraction of the blood, but why this happens is not yet known. They invade the bone marrow, spleen, and liver and there interfere with the normal formation of blood cells. As a result, the patients become anaemic, bleed easily, and repeatedly suffer from a variety of infections: it is one of these that eventually proves fatal.

In the early 1980s, there was no really satisfactory treatment for hairy cell leukemia. It had been found by trial that if the spleen was enlarged, as is usual in this disease, removing it surgically gave benefit to the patient for a time, but did not cure the condition. After a few years, clinical symptoms reappeared and rapidly worsened.

Jordan Gutterman's superior at the M.D. Anderson Hospital was Evan Hersh. This clinician had made a special study of hairy cell leukemia for many years, and had come to the conclusion that the course of the disease and its outlook were greatly influenced by body's own defence mechanisms. He decided to give responsibility for the care of the hairy cell leukemia patients in the hospital, and for carrying out clinical research in relation to this disease, to a young colleague, Jorge Quesada.

Quesada immersed himself in this topic with enthusiasm, and at the same time continued to further his medical education in the different departments of the huge hospital. As a result, he became acquainted with Jordan Gutterman, and the latter soon transmitted to him his own interest in interferon and his faith in its potential. At the beginning of 1982, Quesada had the idea of trying interferon for the treatment of a young man with severe hairy cell leukemia. He discussed this with Gutterman who had to decide how best to make use of the small amounts of precious interferon he was obtaining from Finland. Faced with an almost unlimited number of options, Jordan made the decision to look at hairy cell leukaemia, and in July 1982, Arthur, for that was the name of the patient, was given his first injection of interferon. Within a few weeks the number of platelets in his blood began to increase. This was such a promising sign that Jordan and Jorge decided to enrol other patients in the study. In the course of the following months a number of other amazingly good findings were seen in all the treated patients. Their quality of life improved dramatically; they were no longer plagued by recurring infections, and their blood and bone marrow became more or less normal. Something really significant was happening.

In the following year, Jorge, Jordan, Evan Hersh and their co-workers published an article describing the promising results from treating their first seven hairy cell leukemia patients with Finnish interferon. Three had lost all signs of the leukemia during interferon treatment, while in the remaining four patients there had been a definite though not complete response. The paper appeared in 1984 in the first issue of the internationally read *New England Journal of Medicine,* and as a result, the information spread rapidly around the world. It made a great impact.

In Finland, interferon therapy was first given to patients with hairy cell leukemia in the fall of 1983. Soon after the article of Quesada and his co-workers appeared in print, I began to receive phone calls from doctors in many parts of Finland requesting interferon for the treatment of their patients with hairy cell leukemia. New information published in a leading journal is picked up by the medical community around the world with amazing speed and efficiency.

By the time their first paper appeared in print, the Houston team had already accumulated much additional information about the use of interferon in hairy cell leukemia. It was effective in all stages of the disease and in the majority of the patients, even in those who had not gained benefit from the removal of their spleen. Furthermore, the quality of life improved dramatically for many patients: they were able to go back to work and even to indulge in their previous sports activities. Such results had never been seen before in the treatment of this form of leukaemia.

At this time, the first recombinant interferons were just entering the clinic and Gutterman's team was actively involved. It was at the M.D. Anderson Hospital that such an interferon derived from bacteria was injected for the first time into a human being, and this was only a year after workers in Weissmann's laboratory in Zurich had shown that human interferon could be made from this source. Meanwhile, the competing team from Genentech, led by David Goeddel, had rapidly reached the same goal. The intense competition between the two groups meant that clinical work began at an astonishingly short time after the original scientific studies.

Representatives from the press and television were present on the historic occasion on January 15, 1981, when the first patient received recombinant interferon at the M.D. Anderson Hospital. Seen from the present day, this brash publicity looks rather stupid and ridiculous. The cameramen filmed the patient, and the reporters interviewed him, and then all promptly left. This was perhaps just as well, for after a while, the patient began to develop the same somewhat unpleasant reactions as had been encountered with our own leukocyte interferon during the previous ten years.

These side effects were a considerable disappointment to Gutterman, as to many others. They had been confidently attributed to the impurities in our interferon preparations rather than to interferon itself. These starry-eyed optimists had thought that they would not be encountered when the highly purified interferons from bacteria were used, but there was in fact little ground for such wishful thinking.

The true state of affairs had become clear to me during the previous year. David Secher in Cambridge had utilised his monoclonal antibodies to obtain a virtually completely pure version of our interferon, and Geoff Scott had injected this into volunteers in the Research Centre in Salisbury. In this careful study, other volunteers were given our routinely prepared and only partially purified leucocyte interferon. The two interferon preparations had been coded so that neither the research workers nor the volunteers knew which preparation each person received, and the symptoms and laboratory measurements from each volunteer were recorded carefully. The results showed that the effects of both preparations were indistinguishable. Even pure interferon clearly caused adverse effects. These results were published in *British Medical Journal* early in 1981.

Nevertheless, it was important to see whether the recombinant interferons were as effective as natural leukocyte interferon in the therapy of hairy cell leukemia. After the publication by Quesada and co-workers, doctors from many parts of the U.S. began sending their own patients with hairy cell leukemia to M.D. Anderson Hospital, and a considerable number of patients became available for treatment. Within a year from the appearance of the first paper, 60 patients in total had been treated with recombinant interferon. The results seemed clear-cut: this highly purified recombinant alpha-2 interferon from bacteria was as effective in hairy cell leukaemia as our Finnish leukocyte interferon. Similar results were soon published from other hospitals and other countries, and interferon from Namalwa cells also proved active. The different alpha interferons are without question effective in about 90 per cent of hair cell leukemia patients. In most cases, the interferon treatment does not cure the leukemia, but it keeps it under control in rather the same way as insulin keeps diabetes at bay.

In June 1986, the U.S. Food and Drug Administration approved two recombinant alpha-2 interferons as drugs for the treatment of hairy cell leukemia. By that time, various alpha interferon preparations had already been registered in other countries for use in the same condition. Interferon had at last become an officially approved medicine for the treatment of a form of cancer.

Often, the time needed for the official approval of a new medicine, i.e., its registration as a recognised drug, is lengthy and may take several years. As well as extensive laboratory studies, large and well-controlled clinical trials are normally required, as I have described in the chapter on "Double-blind." Results previously obtained with other treatments, the so-called historical controls, are not usually considered to provide adequate controls. However, interferon therapy was clearly so greatly superior to all the treatment modalities previously tried in hairy cell leukemia, that the control authorities everywhere were fully convinced of its value.

Interferon has existed on this planet for more than a hundred million years, and no doubt members of the human race have been afflicted with hairy cell leukemia for very long. Medical research discovered interferon and identified hairy cell leukemia as a distinct disease at about the same time. Research on these two topics proceeded quite independently for some 25 years and then, and surprisingly, the two lines of research converged. It was found that interferon, a normal protein of the human body, is able to keep the leukemia cells at bay. This is how progress is made in medical research.

The hopes long cherished by me and many others had come true. Although the word "breakthrough" has suffered from overuse, I shall use it again just in this connection, and comment that interferon had now made an indisputable breakthrough. It had been convincingly shown, and for the first time, that a pure natural human protein was an effective treatment for a form of cancer.

Hairy cell leukemia is a rare disease, but the achievement had a great psychological impact on scientists and on the pharmaceutical industry. It inspired confidence that interferon and other similar substances may eventually prove effective even in the treatment of more common forms of malignant disease. Now even those doctors who had been guarded in their attitude toward interferon and overwhelmingly critical had to admit that it did have some practical value. The battered reputation of interferon started on a definite rise.

How does interferon keep hairy cell leukemia under control? Does it inhibit the division of the leukemic cells? Does it change them into normal cells? Does it help the normal defence mechanisms in the body to destroy the leukemic cells? We do not yet have a definite answer to these questions, but equally, the mechanism of action of many other effective medicines is likewise unknown.

45

Finnferon-Alpha

The leukocyte interferon produced by the Finnish Red Cross Blood Transfusion Service was registered as a drug in Finland in 1986. It was first approved for the treatment of laryngeal papillomas and hairy cell leukemia, but other indications have since been added to the list. The product is called Finnferon-alpha, a name chosen after much deliberation in the Blood Centre. The Finnish interferon had often been called "Cantell interferon" in the interferon world, and even such queer names as Cantelferon figured among the candidates; I am glad to say it was rejected in favour of Finnferon.

Finnferon has been a source of great pleasure to me. It is a product based on developmental work carried out over nearly a quarter of a century. It is largely a home-made product which has as its source the blood from voluntary Finnish donors. Even Finnish hens make a contribution, for the virus used for the induction of interferon is grown in their eggs. The State alcohol monopoly also does its bit, providing the ethanol used in its purification.

Leukocyte interferons similar to ours have also been registered elsewhere. My friend Diego Balducci succeeded in registering his product in Italy by the name Alfaferone. It is made and purified like Finnferon, and it sells well. Leukocyte interferons have also been registered in Argentina, China, Hungary, the

United States of America and Yugoslavia, and they are also produced in Denmark, Sweden, Germany, Japan and Russia. As far as I know, leukocyte interferon is made everywhere more or less by the method we have worked out, but the product has been purified to a greater extent in many places. The Finnish Blood Centre has also begun to employ monoclonal antibodies for the complete purification of its interferon, though one of the many interferon components is lost during the process, because it does not bind to these antibodies.

At the same time as Finnferon, a recombinant alpha-2 interferon preparation was registered in Finland. Its commercial name is Intron, and it has its origin in my collaboration with Charles Weissmann. The interferon messenger RNA which I sent to Charles enabled him to get hold of the human interferon genes. Biogen, the company which he and his fellow scientists founded, started the commercial production of one of these gene products, human interferon alpha-2, in bacteria. When Biogen ran into financial difficulties, the established and much bigger firm Schering-Plough acquired it. Thus the big fish ate the small one. Because of the resources of the large organisation, alpha-2 interferon was developed into a drug which entered the Finnish market to compete with our domestically produced Finnferon. Through skilful, effective and bold marketing, Intron took the lion's share of the market for interferon in Finland. Soon, an alpha-2 interferon preparation produced by another large international company, Hoffman La Roche, joined the race, and a bit later this was followed by Wellferon, a product of Namalwa cells; the early history of this product is described in my chapter on "Namalwa." The high interferon-producing capacity of this cell was discovered in the collaboration between Hans Strander, Erik Mogensen and myself in 1974.

Thus I have been actively involved in the development of three tough competitors for Finnferon. So one might say that, from the point of view of our own Finnferon, I have helped to saw through the branch on which I was sitting. But I did not think along these lines. I could see Finnferon's limitations from the very beginning, and realised that the future of interferon depends on the development of methods for producing it in unlimited quantities. Nevertheless, it has caused me some annoyance that so many people considered Finnferon to be obsolete once the recombinant interferons came on the market. Some people have emphasised that Finnferon is impure, and contains only about 1 per cent interferon. This is true, but it is important to keep in mind that this is also true for all its competitors: to improve their stability, albumin, the same blood protein which is present in Finnferon, is included in the final formulation of all the various interferon preparations which are available in pharmacies. In this respect, Finnferon does not dramatically differ from its competitors. Finnferon is not a vague leukocyte extract, and its exact composition is quite well established.

A Festschrift was dedicated to Jean Lindenmann and published to celebrate the 30th anniversary of the discovery of interferon. I contributed an article entitled "Is natural human leukocyte interferon still needed?" In this I pondered whether the time had come for leukocyte interferon to retire from the scene along with its developers. Does our interferon have any advantages over its competitors? In my opinion, it still has a function.

Finnferon contains the 13 different alpha interferons and a considerable amount of omega interferon. Some interferons potentiate each others' actions. Therefore, it is possible that a drug consisting of an interferon cocktail is more effective than a drug containing just one kind of interferon. Virtually no work has been done to compare the clinical performance of different interferons, and this is our present goal. We want to compare the efficacy of Finnferon with that of other interferons in the treatment of human diseases. If Finnferon is no better than its competitors, it is not needed any more. It has fullfilled its job and is ready to go. If Finnferon turns out to be better, then the pharmaceutical industry faces the challenge of developing mixtures of alpha interferons that resemble Finnferon.

The key question is why human cells produce so many different alpha interferons. Do they represent just evolutionary accidents, a profligate waste of resource on the part of the Nature? Would a single alpha interferon be capable of performing the same functions as all the other alpha interferons? Or is it that each different alpha interferon has its own, presently unknown, special function in human body? The latter possibility seems more likely to me, but we do not yet have reliable answers to these questions.

In my opinion, Finnferon is not yet ready to sink into the oblivion of history. It is not impossible that it will prove more effective than its competitors in the treatment of certain diseases. In such a case it will give valuable information to science and industry. For this reason it should also be in the interest of the pharmaceutical industry to compare the efficacy of Finnferon and other interferons in the treatment of various diseases.

The above thoughts represent theoretical speculation, but a practical problem in relation to recombinant interferons has emerged during the past few years. The number of patients treated with interferons has increased and the duration of the interferon therapy has been prolonged. It has been observed that in consequence, a number of patients become immunised against the interferon administered to them, and form antibodies to it. These antibodies bind firmly to the injected interferon and they may totally abolish its effects. It seems that the recombinant alpha interferons are much more prone to evoke the formation of such antibodies than the natural interferons derived from human cells. Why is this?

One obvious explanation is that the structure of the recombinant interferons is subtly different from that of the natural interferons.

The human body notices the difference, and recognises the recombinant interferons as foreign proteins. It therefore tries to reject them by forming the neutralising antibodies. The result is that natural interferon preparations are often still efficacious in patients who no longer respond to recombinant interferons because of antibodies. These observations have opened new indications and new markets for Finnferon and other natural interferons.

46

Interferons As Drugs Today And Tomorrow

During the past years, new drugs have been found which may replace or supplement interferon in the treatment of hairy cell leukemia. At the same time, new and much more common indications have been found for the use of interferons, and their use in medicine has grown and become established. Now interferons are used for the treatment of about ten different malignancies and certain virus infections, and extensive clinical trials in many other conditions are in progress.

On the tumour front, the best results have been obtained in the treatment of malignant blood diseases. In some of these, most of the treated patients have derived benefits from interferon. The results in the treatment of solid tumours have as yet not been as good. Nevertheless, interferon is clearly effective in a proportion of patients with e.g. melanoma, renal cell carcinoma, lymphoma and myeloma. Significantly better results have been obtained when interferon has been combined with other methods of treatment.

On the virus front, alpha interferons are the best agents available for the treatment of chronic infections with hepatitis viruses. These affect hundreds of millions of people around the world, and often lead to liver cirrhosis or cancer. Nearly half of the patients benefit from interferon therapy; ways of increasing its effects are being sought.

Recently beta interferon has been registered as the treatment for a particular form of multiple sclerosis.

Gamma interferon is finding its own uses. It is effective in certain common parasitic infections.

Why is interferon therapy effective only in a proportion of patients? How could its efficacy be improved? How could one identify in advance those patients who will benefit from interferon treatment? Scientists around the world are seeking answers to these essential questions.

In the living body, interferons act together with a number of other regulatory substances. The great potential of interferons will be best exploited by combining them with other means of treatment, and the number of possible combinations is almost unlimited. The work will take centuries. In truth, the interferon story has only just begun.